Representations of Reductive Groups

The representation theory of reductive algebraic groups and related finite reductive groups is a subject of great topical interest and has many applications. The articles in this volume provide introductions to various aspects of the subject, including algebraic groups and Lie algebras, reflection groups, abelian and derived categories, the Deligne–Lusztig representation theory of finite reductive groups, Harish-Chandra theory and its generalisations, quantum groups, subgroup structure of algebraic groups, intersection cohomology, and Lusztig's conjectured character formula for irreducible representations in prime characteristic.

The articles are carefully designed to reinforce one another, and are written by a team of distinguished authors: M. Broué, R. W. Carter, S. Donkin, M. Geck, J. C. Jantzen, B. Keller, M. W. Liebeck, G. Malle, J. C. Rickard, and R. Rouquier. This volume as a whole should provide a very accessible introduction to an important, though technical, subject.

Roger Carter is Professor of Mathematics at the University of Warwick. **Meinolf Geck** holds a research position at the CNRS in Paris.

Publications of the Newton Institute

Edited by H. K. Moffatt
Director, Isaac Newton Institute for Mathematical Sciences

The Isaac Newton Institute of Mathematical Sciences of the University of Cambridge exists to stimulate research in all branches of the mathematical sciences, including pure mathematics, statistics, applied mathematics, theoretical physics, theoretical computer science, mathematical biology and economics. The four six-month long research programmes it runs each year bring together leading mathematical scientists from all over the world to exchange ideas through seminars, teaching and informal interaction.

Representations of Reductive Groups

Edited by

Roger W. Carter

and

Meinolf Geck

CAMBRIDGE
UNIVERSITY PRESS

PUBLISHED BY THE PRESS SYNDICATE OF THE UNIVERSITY OF CAMBRIDGE
The Pitt Building, Trumpington Street, Cambridge CB2 1RP, United Kingdom

CAMBRIDGE UNIVERSITY PRESS
The Edinburgh Building, Cambridge, CB2 2RU, United Kingdom
40 West 20th Street, New York, NY 10011-4211, USA
10 Stamford Road, Oakleigh, Melbourne 3166, Australia

First published 1998

Printed in the United Kingdom at the University Press, Cambridge

A catalogue record for this book is available from the British Library

ISBN 0 521 643252 paperback

Contents

Preface

This volume gives an account of the representation theory of reductive algebraic groups over algebraically closed fields and over finite fields. It contains carefully coordinated chapters written by 9 leading workers in the area of algebraic groups.

The volume begins with an article by R.W. Carter introducing the basic concepts in the theory of linear algebraic groups. This includes the properties of well known subgroups such as maximal tori, Borel subgroups and parabolic subgroups, and a description of the classification of the simple algebraic groups by means of root systems and Dynkin diagrams.

There is a class of abstract groups, the Coxeter groups, which play a key role in the theory of algebraic groups. An article by R. Rouquier discusses the properties of Coxeter groups in general, and also the particular Coxeter groups such as Weyl groups and affine Weyl groups which appear in the theory of algebraic groups.

Various concepts from homological algebra are frequently used in the representation theory of algebraic groups. A chapter by B. Keller introduces these concepts, including abelian categories, derived categories and triangulated categories.

Finite reductive groups are defined as fixed point sets of reductive algebraic group under a Frobenius map. The representation theory in characteristic 0 of these groups was developed by Deligne and Lusztig. An article by M. Geck explains the basic properties of Frobenius maps and expounds the Deligne-Lusztig theory, including a parametrization of all irreducible representations of finite reductive groups.

An elegant theory has been developed by Broué, Malle and Michel in recent years in the context of the modular representation theory of finite reductive groups in the cross characteristic case. There is an exposition of this theory, known as the d–Harish Chandra or generalized Harish–Chandra theory, by M. Broué and G. Malle.

There is also an article by J.C. Jantzen giving an introduction to quantum groups. Quantum groups are now proving useful in many varied branches of mathematics, in particular the modular representation theory of reductive groups in the equal characteristic case. Jantzen explains the main ideas, beginning with quantum sl_2 and leading up to the properties of the canonical basis.

There is an intimate connection between the representation theory of algebraic groups and the subgroup structure of such groups. An article by M.W. Liebeck discusses the subgroup structure of simple algebraic groups both over algebraically closed fields and over finite fields.

Many of the deeper results on representations of algebraic groups, particularly those of G. Lusztig, and various positivity results on Hecke algebras and canonical bases of quantum groups, use methods of intersection cohomology. A chapter by J.C. Rickard provides an introduction to the subject of intersection cohomology.

Properties of intersection cohomology are used in the final article by S. Donkin. There is at present no known character formula for the irreducible modules for reductive groups over fields of prime characteristic. However, there is a conjectural formula due to G. Lusztig, which is known to hold in certain situations. Donkin describes Lusztig's conjectured character formula and what is at present known about it.

Introduction to algebraic groups and Lie algebras

Roger W. Carter

Mathematics Institute, University of Warwick, Coventry CV4 7AL, U.K.

1 Basic concepts

We begin with some basic definitions relating to algebraic groups.

Let k be an algebraically closed field and k^n the vector space of n-tuples over k. An *affine variety* over k is a subset of k^n for some n which is defined by the vanishing of a set of polynomial equations. A *morphism of affine varieties* is a map $\phi : V_1 \to V_2$ such that, for $v \in V_1$, the coordinates of $\phi(v)$ are polynomial functions in the coordinates of v. If $V \subset k^n$ and $V' \subset k^{n'}$ are affine varieties it is clear that $V \times V' \subset k^{n+n'}$ will be an affine variety also.

An *affine algebraic group* over k is a set G which is both an affine variety and a group such that the maps $(x, y) \to xy$ from $G \times G$ to G and $x \to x^{-1}$ from G to G are morphisms of varieties.

For example consider the special linear group given by

$$SL_n(k) = \{(a_{ij}) \in k^{n^2} \ ; \det(a_{ij}) = 1\}.$$

Then $SL_n(k)$ is an affine algebraic group. Its group structure is given by matrix multiplication and its variety structure is given by the fact that it is the set of points in k^{n^2} given by the vanishing of a single polynomial equation.

We note that the same argument does not apply to the general linear group $GL_n(k)$ given by

$$GL_n(k) = \{(a_{ij}) \in k^{n^2} \ ; \det(a_{ij}) \neq 0\}.$$

However $GL_n(k)$ can be given the structure of an affine algebraic group by considering it as a subset of k^{n^2+1} instead. We have

$$GL_n(k) = \{(a_{ij}, b) \in k^{n^2+1} \ ; b \det(a_{ij}) = 1\}.$$

A *homomorphism* of affine algebraic groups is a map $\phi : G_1 \to G_2$ which is a morphism of varieties and a homomorphism of groups. An *isomorphism* of affine algebraic groups is a map $\phi : G_1 \to G_2$ which is bijective such that ϕ and ϕ^{-1} are homomorphisms of affine algebraic groups.

We now consider subvarieties of affine algebraic varieties. If $V \subset k^n$ is an affine variety the subvarieties of V are those subsets of V which are themselves affine varieties in k^n. The subvarieties of V form the closed sets in a topology called the *Zariski topology* on V.

Given an affine algebraic group G it is natural to consider the closed subgroups of G, since such subgroups are themselves affine algebraic groups.

A *linear algebraic group* is a closed subgroup of $GL_n(k)$ for some n. Thus every linear algebraic group is an affine algebraic group. In fact the converse is true also. Given any affine algebraic group G there is an isomorphism of affine algebraic groups between G and a closed subgroup of $GL_n(k)$ for some n. Thus any affine algebraic group is isomorphic to a linear algebraic group. The terms 'affine algebraic group' and 'linear algebraic group' are thus interchangeable. It is more common to use the term 'linear algebraic group'.

We mention three basic reference books on linear algebraic groups where more details and proofs of the results mentioned here can be found. They are:

Linear Algebraic Groups by A. Borel.

Linear Algebraic Groups by J.E. Humphreys.

Linear Algebraic Groups by T.A. Springer.

(See the precise references at the end of this chapter.)

2 Linear algebraic groups

Let G be a linear algebraic group. Then G, considered as an affine variety, will be the union of finitely many irreducible components. As a topological space G will be the disjoint union of its connected components. In fact it can be shown that the irreducible components coincide with the connected components. These will simply be called the components of G. Thus G is the disjoint union of finitely many components. Let G^0 be the component containing the identity element. Then the other components turn out to be simply the cosets of G^0 in G. Furthermore, G^0 is a normal subgroup of G of finite index and the set of cosets G^0x, $x \in G$, is the set of components of G.

An element $x \in G$ is called *semisimple* if x is a diagonalisable matrix, i.e. conjugate in the general linear group to a diagonal matrix. It can be shown that this condition is independent of the matrix representation of G. Thus the property of being semisimple depends only on the linear algebraic group G and not on the particular matrix representation which is used.

An element $x \in G$ is called *unipotent* if all the eigenvalues of the matrix x are equal to 1. This condition is again independent of the matrix representation of G.

It can be shown that each element $x \in G$ is uniquely expressible in the form

$$x = x_s x_u = x_u x_s$$

where x_s is semisimple and x_u is unipotent. This is called the *Jordan decomposition* of x. This decomposition plays a key role in the theory of algebraic groups.

We now define a number of subgroups in a linear algebraic group G.

$R(G)$ is the *radical* of G. This is the unique maximal closed connected soluble normal subgroup of G.

$R_u(G)$ is the *unipotent radical* of G. This is the unique maximal closed connected normal subgroup of G, all of whose elements are unipotent. An algebraic group is called unipotent if all its elements are unipotent. We have

$$R_u(G) \subset R(G).$$

The group G is called *reductive* if $R_u(G) = 1$. It is called *semisimple* if $R(G) = 1$. Thus any semisimple group is reductive, but not conversely. The factor group $G/R_u(G)$ is a reductive group and the factor group $G/R(G)$ is semisimple. In fact the factor group G/N of a linear algebraic group with respect to a closed normal subgroup N has the structure of a linear algebraic group, well defined up to isomorphism.

A linear algebraic group G is called *simple* if G is connected but G has no proper closed connected normal subgroup. Every abelian simple group has dimension 1 (as an affine variety). There are two such groups, the additive group k^+ and the multiplicative group k^*. k^+ can conveniently be defined as

$$k^+ = \left\{ \begin{pmatrix} 1 & \lambda \\ 0 & 1 \end{pmatrix} ; \lambda \in k \right\}$$

since

$$\begin{pmatrix} 1 & \lambda \\ 0 & 1 \end{pmatrix} \begin{pmatrix} 1 & \mu \\ 0 & 1 \end{pmatrix} = \begin{pmatrix} 1 & \lambda + \mu \\ 0 & 1 \end{pmatrix}.$$

The multiplicative group k^* is defined by $k^* = GL_1(k)$ since

$$(\lambda)(\mu) = (\lambda\mu) \qquad \lambda \neq 0, \mu \neq 0.$$

If G is a non-abelian simple algebraic group then G is semisimple. The centre $Z(G)$ of such a group is finite, and the factor group $G/Z(G)$ is simple as an abstract group. G itself need not be simple as an abstract group. For example $SL_2(k)$ is simple as an algebraic group but not necessarily as an abstract group. For this reason the algebraic groups we have defined as simple are sometimes called almost simple groups.

3 Maximal tori and Borel subgroups

It is possible to define the direct product of linear algebraic groups in a natural way. In particular, an algebraic group isomorphic to a direct product

$$k^* \times \cdots \times k^*$$

of multiplicative groups is called a *torus.* A maximal torus in a linear algebraic group G is a closed subgroup of G which is a torus but is not contained in any larger torus. It can be shown that any two maximal tori are conjugate in G.

Let G be a connected linear algebraic group and T be a maximal torus of G. Let $C(T)$ be the centralizer of T and $N(T)$ be the normalizer of T. Then we have

$$T \subset C(T) \subset N(T).$$

In fact one can show that $C(T) = N(T)^0$, the connected component of $N(T)$ containing 1. In particular $N(T)/C(T)$ is finite. We write

$$W = N(T)/C(T).$$

W is called the *Weyl group* of G. It is well defined by G up to isomorphism, since any two maximal tori of G are conjugate. In the special case when G is a connected reductive group we have $T = C(T)$. Thus in this case we have

$$W = N(T)/T.$$

The Weyl group W is a finite Coxeter group, i.e. it is a finite group which can be defined by generators and relations of the form

$$\begin{aligned} W = \langle s_1, \dots, s_n; \qquad & s_i^2 = 1 \qquad \text{for } i = 1, \dots, n \\ & (s_i s_j)^{m_{ij}} = 1 \quad \text{for } i \neq j \rangle. \end{aligned}$$

Suppose, for example, that $G = GL_n(k)$. This is a connected reductive group. The subgroup $D_n(k)$ of diagonal matrices is a maximal torus T of G. The normalizer $N(T)$ consists of all monomial matrices in G. Thus we have

$$N(T)/T \cong S_n$$

where S_n is the symmetric group of degree n. Thus the Weyl group W is isomorphic to S_n. Consider the transpositions of consecutive integers given by

$$s_1 = (1,2), \quad s_2 = (2,3), \quad \dots, \quad s_{n-1} = (n-1, n).$$

Then we have

$$\begin{aligned} s_i^2 &= 1 && \text{for } i = 1, \ldots, n-1. \\ (s_i s_j)^2 &= 1 && \text{for } |i-j| > 1. \\ (s_i s_j)^3 &= 1 && \text{for } |i-j| = 1. \end{aligned}$$

Moreover $s_1, \ldots, s_{n-1}$ generate $S_n \cong W$ as a Coxeter group.

We now consider a type of subgroup, introduced by A. Borel, which plays a key role in the theory of algebraic groups. A *Borel subgroup* of a linear algebraic group G is a maximal closed connected soluble subgroup of G. (In fact a maximal connected soluble subgroup is always closed.) It can be shown that any two Borel subgroups of G are conjugate. This is proved by showing that, if B is a Borel subgroup of G, then the set G/B of left cosets can be regarded as a projective variety and that a connected soluble group, when acting on a complete variety (in particular on a projective variety) always has a fixed point. This illustrates the fact that in order to develop the theory of linear algebraic groups one cannot remain within the category of affine varieties. Additional types of algebraic varieties, such as projective varieties, must be used also.

Let G be a connected linear algebraic group and B be a Borel subgroup of G. Then it can be shown, by a somewhat subtle argument, that B is its own normalizer in G. Now any maximal torus T of G, being a closed connected soluble subgroup, lies in some Borel subgroup B of G. Let $U = R_u(B)$ be the unipotent radical of B. Then we have

$$B = UT \quad \text{and} \quad U \cap T = 1.$$

Thus B is a semidirect product of the unipotent group U and the torus T.

For example, let $G = GL_n(k)$. Then G is a connected reductive group. The diagonal subgroup $T = D_n(k)$ is a maximal torus in G. The subgroup $B = T_n(k)$ of upper-triangular matrices is a Borel subgroup of G and the subgroup $U = U_n(k)$ of upper unitriangular matrices is the unipotent radical of B.

Let $N = N(T)$. Then it can be shown that

$$G = BNB$$

i.e. each double coset of B in G contains an element of N, so has form BnB for some $n \in N$. Since $W = N/T$ we have a natural homomorphism $\pi : N \to W$. Then we have, for $n, n' \in N$, $BnB = Bn'B$ if and only if $\pi(n) = \pi(n')$. This gives a bijective correspondence between the set $B\backslash G/B$ of double cosets of B in G and the set of elements of W. The factorisation $G = BNB$ is called the *Bruhat decomposition* of G.

It is a remarkable fact that the set of all subgroups of G containing B can be described in a simple way. Let $I = \{1, 2, \ldots, n\}$ and $W = \langle s_1, \ldots, s_n \rangle$ be a set of Coxeter generators for the Weyl group W. Let $J \subset I$ be any subset of I. Let W_J be the subgroup of W given by

$$W_J = \langle s_i \, ; \, i \in J \rangle .$$

Let N_J be the subgroup of N satisfying

$$N_J / T = W_J .$$

Then it can be shown that

$$P_J = B N_J B$$

is a subgroup of G containing B, and that any subgroup of G containing B is equal to P_J for some $J \subset I$. Furthermore $P_J = P_{J'}$ if and only if $J = J'$. We also have

$$\begin{aligned} \langle P_J \, , P_{J'} \rangle &= P_{J \cup J'} \\ P_J \cap P_{J'} &= P_{J \cap J'} \end{aligned}$$

thus there are 2^n subgroups of G containing B which form a lattice isomorphic to the lattice of subsets of the set I with n elements. A *parabolic subgroup* of G is a subgroup conjugate to P_J for some J. A subgroup of G is parabolic if and only if it contains a Borel subgroup of G. Each parabolic subgroup can be shown to be its own normalizer in G.

For example, let $G = GL_n(k)$. The subgroup $B = T_n(k)$ of upper triangular matrices in G is a Borel subgroup. The parabolic subgroups containing B are the 'staircase subgroups' of the form

$$P_J = \begin{pmatrix} \boxed{*} & & & * \\ & \boxed{*} & & \\ & & \boxed{*} & \\ 0 & & & \boxed{*} \end{pmatrix}$$

where the matrices in the diagonal blocks are non-singular and the entries in the blocks above the diagonal are arbitrary. The number of such staircase subgroups is 2^{n-1}, which relates to the fact that the Weyl group $W = S_n$ has $n-1$ Coxeter generators.

Returning to the general case let $U_J = R_u(P_J)$. Then there exists a closed subgroup L_J of P_J such that

$$P_J = U_J L_J \, , \qquad U_J \cap L_J = 1 \, .$$

L_J is called a Levi subgroup of P_J. L_J is a connected reductive group. Thus the parabolic subgroup P_J is a semidirect product of its unipotent radical with a Levi subgroup. Also any two Levi subgroups of P_J are conjugate in P_J.

For example in the parabolic subgroup P_J of $GL_n(k)$ given above we may take

$$U_J = \begin{pmatrix} \boxed{I} & & & * \\ & \boxed{I} & & \\ & & \boxed{I} & \\ 0 & & & \boxed{I} \end{pmatrix} \qquad L_J = \begin{pmatrix} \boxed{*} & & & 0 \\ & \boxed{*} & & \\ & & \boxed{*} & \\ 0 & & & \boxed{*} \end{pmatrix}.$$

We note that the Levi subgroup L_J is a connected reductive group containing T as a maximal torus whose Weyl group is

$$N_{L_J}(T)/T \cong W_J\,.$$

4 Roots and coroots

We shall now describe the root system and coroot system of a connected reductive group G. The root system lies in the character group of a maximal torus T of G and the coroot system lies in the cocharacter group of T, so we begin by defining the character and cocharacter groups of a torus.

Let T be a torus and $X = \mathrm{Hom}(T, k^*)$ be the set of homomorphisms of algebraic groups from T to the multiplicative group k^*. X has a natural structure of an additive group under the operation

$$(\chi_1 + \chi_2)t = \chi_1(t)\chi_2(t) \qquad \chi_1\,, \chi_2 \in X\,,\ t \in T\,.$$

In the special case when $T \cong k^*$ we have

$$\mathrm{Hom}(k^*\,, k^*) \cong \mathbb{Z}$$

since the only algebraic group homomorphisms from k^* into itself are the maps $\lambda \to \lambda^m$ for $m \in \mathbb{Z}$. It follows that if

$$T \cong k^* \times \cdots \times k^* \qquad (n \text{ factors})$$

then

$$X \cong \mathbb{Z} \oplus \cdots \oplus \mathbb{Z} \qquad (n \text{ factors})\,.$$

X is called the character group of T.

Now let $Y = \mathrm{Hom}(k^*, T)$. Then we have

$$Y \cong \mathbb{Z} \oplus \cdots \oplus \mathbb{Z} \qquad (n \text{ factors})$$

and Y is called the cocharacter group of T. The group operation on Y is given by

$$(\gamma_1 + \gamma_2)\lambda = \gamma_1(\lambda)\gamma_2(\lambda) \qquad \gamma_1, \gamma_2 \in Y\,,\ \lambda \in k^*\,.$$

There is a duality map

$$X \times Y \to \mathbb{Z}$$

relating the character and cocharacter groups of T. Given $\chi \in X$ and $\gamma \in Y$ we have $\chi \circ \gamma \in \mathrm{Hom}(k^*, k^*)$, thus $(\chi \circ \gamma)\lambda = \lambda^m$ for some $m \in \mathbb{Z}$. We write $\langle \chi, \gamma \rangle = m$ and then $\langle \chi, \gamma \rangle$ is the required duality map. It induces isomorphisms

$$X \cong \mathrm{Hom}(Y, \mathbb{Z}), \qquad Y \cong \mathrm{Hom}(X, \mathbb{Z}).$$

Now let T be a maximal torus in a connected reductive group G. Let $N = N(T)$ and $N/T = W$. Then W acts on T by conjugation, giving an action $t \to t^w$ for $t \in T$, $w \in W$. We may also define actions of W on X and Y by

$$\begin{aligned} {}^w\chi(t) &= \chi(t^w) \text{ for } \chi \in X, t \in T, w \in W \\ \gamma^w(\lambda) &= \gamma(\lambda)^w \text{ for } \gamma \in Y, \lambda \in k^*, w \in W . \end{aligned}$$

Let B be a Borel subgroup of G containing T. It can be shown that there is a unique opposite Borel subgroup B^- such that $B \cap B^- = T$. We have $B = UT$, $B^- = U^-T$ where $U = R_u(B)$ and $U^- = R_u(B^-)$. The torus T acts on U and U^- by conjugation.

We now consider the minimal proper subgroups of U invariant under T. Each of these turns out to be isomorphic to the additive group k^+. T acts on each such subgroup by conjugation, thus giving a homomorphism

$$T \to \mathrm{Aut}\ k^+$$

where Aut k^+ is the group of all algebraic group automorphisms of k^+. Now the only algebraic group automorphisms of k^+ are the maps

$$\lambda \to \mu\lambda \qquad \mu \in k^*.$$

It follows that

$$\mathrm{Aut}\ k^+ \cong k^* .$$

Thus we obtain a homomorphism $T \to k^*$, i.e. an element of the character group X of T. In this way each minimal T-invariant subgroup of U gives an element $\alpha \in X$. Distinct subgroups of U give distinct elements of X. Let Φ^+ be the set of all elements of X arising in this way. Φ^+ is a finite subset of X called the set of positive roots. Each positive root $\alpha \in \Phi^+$ arises from a root subgroup $U_\alpha \subset U$.

Similarly we may consider the minimal T-invariant subgroups of U^-. These give a set of elements Φ^- in X called the set of negative roots. We have

$$\alpha \in \Phi^+ \text{ if and only if } -\alpha \in \Phi^- .$$

The set $\Phi = \Phi^+ \cup \Phi^-$ is called the set of roots of G with respect to T.

We illustrate this idea in the example $G = GL_n(k)$. Let $T = D_n(k)$ be the diagonal subgroup and $B = T_n(k)$, $B^- = T_n^-(k)$ be the subgroups of upper triangular and lower triangular matrices in G. Then the positive root subgroups U_α, $\alpha \in \Phi^+$, have the form

$$\begin{pmatrix} 1 & & \\ & \ddots & \lambda \\ & & \ddots & \\ & & & 1 \end{pmatrix} = I + \lambda E_{ij}$$

for fixed i, j with $i < j$, and where $\lambda \in k$ is in the (i, j)-position. The root α coming from this subgroup is the character

$$\begin{pmatrix} \lambda_1 & & & \\ & \ddots & & \\ & & \ddots & \\ & & & \lambda_n \end{pmatrix} \xrightarrow{\alpha} \lambda_i \lambda_j^{-1} \qquad i < j\,.$$

Similarly the negative root subgroups U_α, $\alpha \in \Phi^-$, have the form

$$\begin{pmatrix} 1 & & & \\ & \ddots & & \\ & \lambda & \ddots & \\ & & & 1 \end{pmatrix} = I + \lambda E_{ij}$$

for fixed i, j with $i > j$ and where $\lambda \in k$ is in the (i, j)-position. The root α coming from this subgroup is the character

$$\begin{pmatrix} \lambda_1 & & & \\ & \ddots & & \\ & & \ddots & \\ & & & \lambda_n \end{pmatrix} \xrightarrow{\alpha} \lambda_i \lambda_j^{-1} \qquad i > j\,.$$

We now return to the case of a general connected reductive group G. Let $\alpha \in \Phi$. Then $-\alpha \in \Phi$ also, and we have root subgroups U_α, $U_{-\alpha}$ of G. Consider the subgroup $\langle U_\alpha, U_{-\alpha}\rangle$ generated by $U_\alpha, U_{-\alpha}$. It can be shown that this subgroup is isomorphic to $SL_2(K)$ or $PSL_2(K) = SL_2(K)/Z$ and that there is a surjective homomorphism

$$SL_2(K) \xrightarrow{\phi} \langle U_\alpha, U_{-\alpha}\rangle$$

satisfying

$$\left\{\phi\begin{pmatrix}1 & *\\ 0 & 1\end{pmatrix}\right\} = U_\alpha$$

$$\left\{\phi\begin{pmatrix}1 & 0\\ * & 1\end{pmatrix}\right\} = U_{-\alpha}$$

$$\left\{\phi\begin{pmatrix}\lambda & 0\\ 0 & \lambda^{-1}\end{pmatrix}\right\} = T.$$

Thus there is a homomorphism $k^* \xrightarrow{\alpha^v} T$ given by

$$\alpha^v(\lambda) = \phi\begin{pmatrix}\lambda & 0\\ 0 & \lambda^{-1}\end{pmatrix}.$$

The element $\alpha^v \in \mathrm{Hom}(k^*, T) = Y$ is called the coroot of α. The root α and its coroot α^v are related by $\langle\alpha, \alpha^v\rangle = 2$. The set of all coroots α^v for $\alpha \in \Phi$ is called Φ^v. Φ^v is a finite subset of Y.

We have seen that the Weyl group W acts on both X and Y. We can now say more about this action. Let

$$n_\alpha = \phi\begin{pmatrix}0 & 1\\ -1 & 0\end{pmatrix} \in \langle U_\alpha, U_{-\alpha}\rangle.$$

Then n_α lies in $N(T)$, and since $W = N(T)/T$, the element n_α induces an element $s_\alpha \in W$. It can be shown that $s_\alpha = s_{-\alpha}$, $s_\alpha^2 = 1$ and that the set of all s_α for $\alpha \in \Phi$ generate the Weyl group W.

The element s_α acts on the character group X of T by

$$s_\alpha(\chi) = \chi - \langle\chi, \alpha^v\rangle\alpha \qquad \chi \in X$$

and on the cocharacter group Y of T by

$$s_\alpha(\gamma) = \gamma - \langle\alpha, \gamma\rangle\alpha^v \qquad \gamma \in Y.$$

Moreover $s_\alpha(\Phi) = \Phi$ and $s_\alpha(\Phi^v) = \Phi^v$ for all $\alpha \in \Phi$. Since the $s_\alpha, \alpha \in \Phi$, generate W it follows that

$$w(\Phi) = \Phi \qquad w(\Phi^v) = \Phi^v$$

for all $w \in W$. Thus the elements of the Weyl group permute the roots and also permute the coroots.

5 Classification of simple algebraic groups

Let G be a simple algebraic group over k, T be a maximal torus of G, and B a Borel subgroup of G containing T. Let Φ be the root system of G with respect to T and Φ^+ the set of positive roots with respect to B. Let Δ be the subset of Φ^+ defined by

$$\Delta = \left\{\alpha \in \Phi^+; \alpha \neq \beta + \lambda \text{ for } \beta, \lambda \in \Phi^+\right\}.$$

Δ is called the set of simple roots, or fundamental roots in Φ^+. Let $\Delta = \{\alpha_1, \ldots, \alpha_n\}$. Then the simple roots are linearly independent elements of the vector space $V = X \otimes_{\mathbb{Z}} \mathbb{R}$ and each root in Φ^+ has form

$$\alpha = c_1\alpha_1 + \cdots + c_n\alpha_n \qquad c_i \in \mathbb{Z},\ c_i \geqslant 0.$$

Each simple root α_i gives rise to an involution s_{α_i} in the Weyl group. We write $s_i = s_{\alpha_i}$. Then the elements $s_1, \ldots, s_n$ generate W as a Coxeter group. Thus W has a presentation

$$W = \langle s_1, \ldots, s_n\ ;\ s_i^2 = 1 \quad (s_is_j)^{m_{ij}} = 1\rangle$$

where m_{ij} is the order of s_is_j.

Now $\alpha_j \in X$ and $\alpha_i^v \in Y$ and so $\langle \alpha_j, \alpha_i^v\rangle \in \mathbb{Z}$ for all $i, j \in \{1, \ldots, n\}$. We define $A_{ij} = \langle \alpha_j, \alpha_i^v\rangle$ and let A be the $n \times n$ matrix $A = (A_{ij})$. The integers A_{ij} are called the *Cartan integers* and A is called the *Cartan matrix*. We have

$$\begin{aligned} A_{ii} &= 2 \qquad i = 1, \ldots, n \\ A_{ij} &\in \{0, -1, -2, -3\} \qquad i \neq j. \end{aligned}$$

Let $n_{ij} = A_{ij}A_{ji}$ for $i \neq j$. It can be shown that

$$n_{ij} \in \{0, 1, 2, 3\} \qquad \text{for all } i \neq j.$$

The numbers n_{ij} determine the numbers m_{ij} which appear in the presentation of W as a Coxeter group. We have:

$$\begin{aligned} n_{ij} = 0 &\implies m_{ij} = 2 \\ n_{ij} = 1 &\implies m_{ij} = 3 \\ n_{ij} = 2 &\implies m_{ij} = 4 \\ n_{ij} = 3 &\implies m_{ij} = 6. \end{aligned}$$

We now describe a geometric configuration associated with this situation. The character group X of T is a free abelian group of rank n and $V = X \otimes_{\mathbb{Z}} \mathbb{R}$ is a vector space over $\mathbb{R}$ of dimension n with basis $\alpha_1, \ldots, \alpha_n$. The action of W on X described above extends to a W-action on V which makes V into an irreducible W-module. There is then a W-invariant positive definite quadratic form on V, unique up to a positive scalar. This makes V into a Euclidean space. The Weyl group W acts on V as a group of isometries, and the element $s_i \in W$ acts on V as the reflection in the hyperplane orthogonal to the simple root α_i. Thus W is a finite group generated by reflections.

We now define the Dynkin diagram. This is a graph with n vertices corresponding to the simple roots $\alpha_1, \ldots, \alpha_n$. Vertices i and j are joined by n_{ij} edges if $i \neq j$. Moreover if $n_{ij} = 2$ or 3 we attach an arrow pointing from i to j when $|\alpha_i| > |\alpha_j|$. In fact if $n_{ij} = 2$ we have $|\alpha_i| = \sqrt{2}|\alpha_j|$ and if $n_{ij} = 3$, $|\alpha_i| = \sqrt{3}|\alpha_j|$. In the case $n_{ij} = 1$ the roots α_i, α_j have the same length.

The Dynkin diagram is uniquely determined by the simple algebraic group G, being independent of the various choices used to define it. The Dynkin diagram is always connected when the algebraic group G is simple.

We now describe the possible Dynkin diagrams which can arise. They are given in the following list.

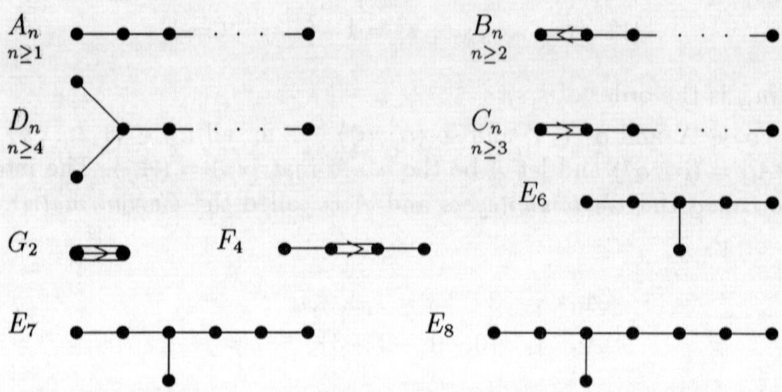

Each simple algebraic group determines one of these Dynkin diagrams, and each Dynkin diagram arises in this way. A given Dynkin diagram can arise from more than one simple algebraic group, but the simple algebraic groups with a given Dynkin diagram are closely related.

A simple algebraic group is said to be of *adjoint type* if $X = \mathbb{Z}\Phi$, i.e. the roots generate the character group. A simple group is called *simply-connected* if $Y = \mathbb{Z}\Phi^v$, i.e. the coroots generate the cocharacter group. There is a unique simple group of adjoint type with a given Dynkin diagram and also a unique simple group of simply-connected type. There are also simple algebraic groups which are neither adjoint nor simply-connected. However the abstract simple group $G/Z(G)$ is the same, up to isomorphism, for all simple algebraic groups G with a given Dynkin diagram.

We shall now describe the configurations formed by the root systems of simple algebraic groups in the Euclidean space of dimension n in the cases when $n = 1$ and $n = 2$. The possible Dynkin diagrams are A_1, A_2, B_2, G_2.

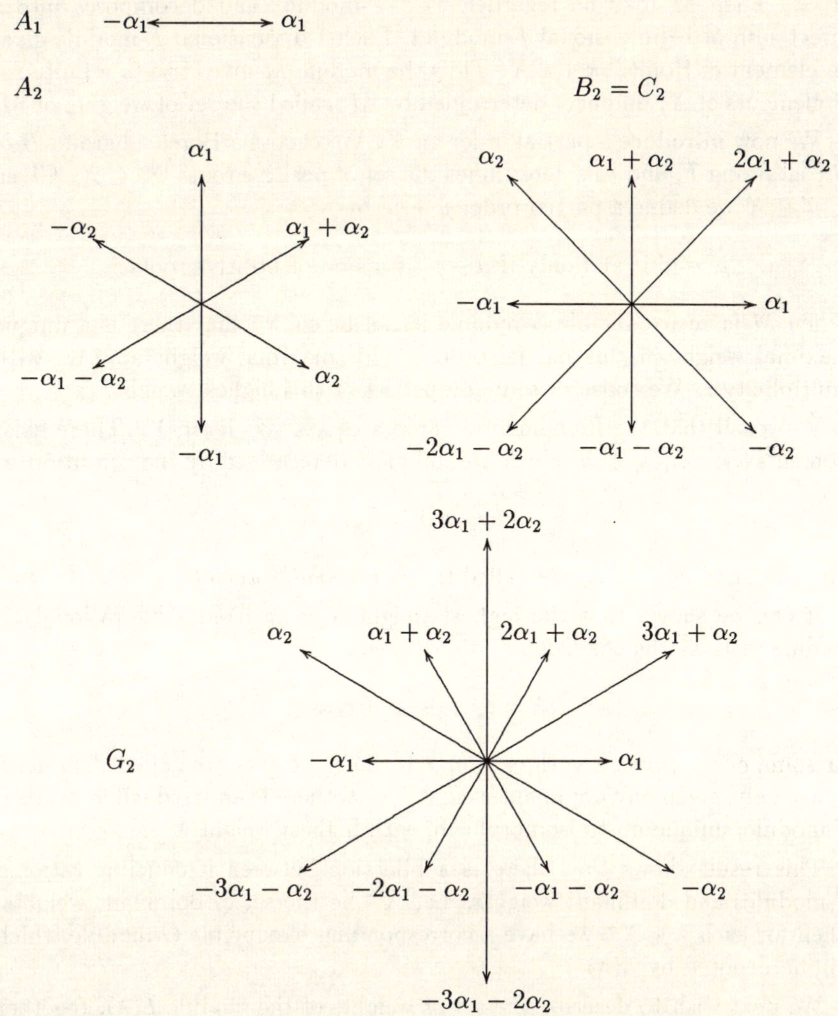

6 Representations of simple algebraic groups

Let G be a simple simply-connected algebraic group over k. A *rational representation* of G is a homomorphism of algebraic groups

$$\rho : G \to GL_n(k)$$

for some n. Such a representation comes from a rational G-module of dimension n. We consider irreducible modules, i.e. those with no proper submodules, since it is known that every rational G-module is a direct sum of irreducible submodules.

So let M be an irreducible rational G-module. Let T be a maximal torus of G. Then M may be regarded as a T-module, and decomposes into a direct sum of 1-dimensional T-modules. Each 1-dimensional T-module gives an element of $\mathrm{Hom}(T, k^*) = X$. Thus the module M gives rise to a finite set of elements of X, uniquely determined by M, called the set of weights of M.

We now introduce a partial order on X. We choose a Borel subgroup B of G containing T, and this determines the set of positive roots $\Phi^+ \subset X$. Given μ, $\mu' \in X$ we define a partial order $\mu \succ \mu'$ by

$$\mu \succ \mu' \text{ if and only if } \mu - \mu' \text{ is a sum of positive roots}.$$

When M is an irreducible G-module it can be shown that there is a unique maximal weight in this partial order. This maximal weight appears with multiplicity 1. We consider some properties of this highest weight.

We recall that the fundamental coroots $\alpha_1^v, \ldots, \alpha_n^v$ lie in Y. There exist elements $\omega_1, \ldots, \omega_n \in X$ which are uniquely determined by the condition

$$\langle \omega_i \, , \alpha_j^v \rangle = \delta_{ij} \, .$$

The elements $\omega_1, \ldots, \omega_n$ are called the *fundamental weights.*

It can be shown that the highest weight λ of an irreducible rational G-module satisfies the condition

$$\lambda = c_1\omega_1 + \cdots + c_n\omega_n$$

for some $c_1, \ldots, c_n \in \mathbb{Z}$ with each $c_i \geqslant 0$. Such weights are called *dominant.* Conversely, given any dominant weight $\lambda \in X$ there is an irreducible rational G-module, unique up to isomorphism, with highest weight λ.

This result shows that there is a bijection between irreducible rational G-modules and dominant weights. Let X^+ be the set of dominant weights. Then for each $\lambda \in X^+$ we have a corresponding irreducible G-module which will be denoted by $L(\lambda)$.

We next wish to describe the set of weights of the module $L(\lambda)$, together with their multiplicities. This information is provided by the character of the module, which is defined as follows.

Let $e(X)$ be a multiplicative group isomorphic to the additive group X and let $\mu \to e^\mu$ be an isomorphism from X to $e(X)$. Then we have

$$e^{\mu_1+\mu_2} = e^{\mu_1} e^{\mu_2} \, . \qquad \mu_1 \, , \mu_2 \in X \, .$$

Let $\mathbb{Z}e(X)$ be the integral group ring of $e(X)$. This is an integral domain, and so has a field of fractions.

For each rational G-module M we define the character of M by

$$\mathrm{ch} M = \sum m_\mu e^\mu \in \mathbb{Z}e(X)$$

where m_μ is the multiplicity of μ as a weight of M. There is a famous formula of H. Weyl which determines the characters of the irreducible G-modules $L(\lambda)$, $\lambda \in X^+$, where G is a simple, simply-connected algebraic group over $\mathbb{C}$. In order to describe Weyl's character formula we introduce some notation. There is a homomorphism

$$\epsilon : W \to \{1, -1\}$$

uniquely determined by $\epsilon(s_i) = -1$ for $i = 1, \ldots, n$. ϵ is called the sign character of the Weyl group. We also define an element $\rho \in X$ given by

$$\rho = \omega_1 + \cdots + \omega_n .$$

(An alternative expression for ρ is

$$\rho = \frac{1}{2} \sum_{\alpha \in \Phi^+} \alpha .)$$

Then, for each $\lambda \in X^+$, we have Weyl's character formula

$$\mathrm{ch} L(\lambda) = \frac{\sum_{w \in W} \epsilon(w) e^{w(\lambda+\rho)}}{\sum_{w \in W} \epsilon(w) e^{w(\rho)}} .$$

This is an equality in the field of fractions of $\mathbb{Z}e(X)$.

We give an example to illustrate Weyl's character formula. Let G have type A_1. Then there is a single fundamental weight ω_1 and so $\lambda = m\omega_1$ for some $m \geqslant 0$. We have $\rho = \omega_1$ and so $\lambda + \rho = (m+1)\omega_1$. The Weyl group W is given by $W = \{1, s_1\}$ and we have $s_1(\omega_1) = -\omega_1$. Thus Weyl's character formula gives

$$\begin{aligned} \mathrm{ch} L(\lambda) &= \frac{e^{\lambda+\rho} - e^{s_1(\lambda+\rho)}}{e^{\rho} - e^{s_1(\rho)}} = \frac{e^{(m+1)\omega_1} - e^{-(m+1)\omega_1}}{e^{\omega_1} - e^{-\omega_1}} \\ &= e^{m\omega_1} + e^{(m-2)\omega_1} + \cdots + e^{-m\omega_1} . \end{aligned}$$

Thus the weights of $L(\lambda)$ are

$$m\omega_1 , \ (m-2)\omega_1 , \ldots , -m\omega_1$$

each with multiplicity 1.

There is also a formula for $\dim L(\lambda)$ due to H. Weyl, called Weyl's dimension formula. Some special cases are:

A_1
$$\begin{aligned} \lambda &= m\omega_1 . \\ \dim L(\lambda) &= m+1 . \end{aligned}$$

A_2
$$\begin{aligned} \lambda &= m_1\omega_1 + m_2\omega_2 . \\ \dim L(\lambda) &= \tfrac{1}{2}(m_1+1)(m_2+1)(m_1+m_2+2) . \end{aligned}$$

In general there is one factor in the numerator for each positive root.

We remark that Weyl's character and dimension formulae are valid for simple, simply-connected algebraic groups over an algebraically closed field of characteristic 0. In the case when k is algebraically closed of prime characteristic p, Weyl's character formula no longer holds. The character of $L(\lambda)$ is not known in this case. However there is a conjectured formula for $\mathrm{ch}L(\lambda)$ if p is not too small, due to G. Lusztig. Lusztig's conjectured character formula will be discussed in a subsequent chapter of this book.

7 The Lie algebra of a linear algebraic group

Let G be a linear algebraic group and $T_1(G)$ be the tangent space of the affine variety G at the identity element. Because G has a group structure it can be shown that $T_1(G)$ has a Lie algebra structure, i.e. given $x, y \in T_1(G)$ there exists $[xy] \in T_1(G)$ satisfying:

$$[xy] \text{ is linear in } x \text{ and in } y$$
$$[xx] = 0$$
$$[[xy]\,z] + [[yz]\,x] + [[zx]\,y] = 0.$$

We denote this Lie algebra by $\mathrm{Lie}(G)$.

Given a homomorphism $\phi : G \to H$ of algebraic groups its differential $d\phi : \mathrm{Lie}(G) \to \mathrm{Lie}(H)$ is a homomorphism of Lie algebras, i.e. a linear map preserving the Lie multiplication.

For each $x \in G$ we have an inner automorphism $i_x : G \to G$ given by $i_x(g) = xgx^{-1}$. We define $\mathrm{Ad}x : \mathrm{Lie}(G) \to \mathrm{Lie}(G)$ by $\mathrm{Ad}x = d(i_x)$. $\mathrm{Ad}x$ is an automorphism of $\mathrm{Lie}(G)$. In fact we have a homomorphism

$$G \to \mathrm{Aut}\ \mathrm{Lie}(G)$$

given by $x \to \mathrm{Ad}x$.

We now suppose that G is a connected reductive group. Let T be a maximal torus of G and B a Borel subgroup of G containing T. Let $\Phi = \Phi^+ \cup \Phi^-$ be the set of roots of G with respect to T and U_α, $\alpha \in \Phi$, be the corresponding set of root subgroups. We write

$$\mathfrak{g} = \mathrm{Lie}(G)\,,\ \mathfrak{h} = \mathrm{Lie}(T)\,,\ \mathfrak{x}_\alpha = \mathrm{Lie}(U_\alpha)\,.$$

Then $\mathfrak{h}$ and $\mathfrak{x}_\alpha$ can be identified with Lie subalgebras of $\mathfrak{g}$ using the differentials of the inclusion homomorphisms in G. We have

$$\mathfrak{g} = \mathfrak{h} \oplus \sum_{\alpha\in\Phi} \mathfrak{x}_\alpha\,.$$

This direct sum decomposition is called the *Cartan decomposition* of $\mathfrak{g}$ with respect to $\mathfrak{h}$, and $\mathfrak{h}$ is called a *Cartan subalgebra* of $\mathfrak{g}$. Each subalgebra $\mathfrak{x}_\alpha$ has dimension 1.

We may also take together the $\mathfrak{x}_\alpha$ for all positive roots, and for all negative roots. Let

$$\mathfrak{n} = \bigoplus_{\alpha\in\Phi^+} \mathfrak{x}_\alpha, \qquad \mathfrak{n}^- = \bigoplus_{\alpha\in\Phi^-} \mathfrak{x}_\alpha.$$

Then $\mathfrak{n}$ and $\mathfrak{n}^-$ are subalgebras of $\mathfrak{g}$ and we have

$$\mathfrak{g} = \mathfrak{n} \oplus \mathfrak{h} \oplus \mathfrak{n}^- .$$

This is called the *triangular decomposition* of $\mathfrak{g}$. We define $\mathfrak{b} = \mathfrak{n} \oplus \mathfrak{h}$. Then $\mathfrak{b}$ is a subalgebra of $\mathfrak{g}$ which is the Lie algebra of B.

For example, suppose $G = GL_n(k)$. Then $\mathfrak{g} = \mathrm{Lie}(G)$ is isomorphic to the Lie algebra of all $n \times n$ matrices over k under Lie multiplication

$$[A, B] = AB - BA .$$

If we take T to be the diagonal subgroup of $GL_n(k)$ then $\mathfrak{h}$ is the subalgebra of $\mathfrak{g}$ consisting of all diagonal matrices. $\mathfrak{n}$ may be taken as the subalgebra of all upper triangular matrices with zeros on the diagonal, and $\mathfrak{n}^-$ as the corresponding subalgebra of lower zero-triangular matrices.

Next suppose that G is a simple, simply-connected algebraic group over $\mathbb{C}$. Then $\mathfrak{g} = \mathrm{Lie}(G)$ is a simple Lie algebra over $\mathbb{C}$ and all finite dimensional simple Lie algebras over $\mathbb{C}$ arise in this way. The finite dimensional simple Lie algebras over $\mathbb{C}$ are thus in bijective correspondence with the Dynkin diagrams.

Every rational representation

$$\rho : G \to GL_n(\mathbb{C})$$

gives rise to a representation of $\mathfrak{g}$

$$d\rho : \mathfrak{g} \to \mathrm{Lie}(GL_n(\mathbb{C})) = [M_n(\mathbb{C})] .$$

$d\rho$ is a homomorphism of Lie algebras from g into the Lie algebra $[M_n(\mathbb{C})]$ of all $n \times n$ matrices under Lie multiplication. Each irreducible representation ρ of G gives rise to an irreducible representation $d\rho$ of $\mathfrak{g}$, and all finite dimensional irreducible representations of $\mathfrak{g}$ arise in this way. Thus the finite dimensional irreducible representations of $\mathfrak{g}$ are in bijective correspondence with the dominant weights, and their characters and dimensions are given by Weyl's character and dimension formulae.

There is an infinite dimensional associative algebra $\mathcal{U}(\mathfrak{g})$ whose representation theory is the same as that of the finite dimensional Lie algebra $\mathfrak{g}$. $\mathcal{U}(\mathfrak{g})$ is called the *universal enveloping algebra* of $\mathfrak{g}$ and is the unique associative algebra with 1 satisfying the following conditions.

(a) There is a linear map $i : \mathfrak{g} \to \mathcal{U}(\mathfrak{g})$ satisfying

$$i[xy] = i(x)i(y) - i(y)i(x) \quad \text{for all } x, y \in \mathfrak{g}.$$

(b) For any associative algebra A with 1 and any linear map $j : \mathfrak{g} \to \mathcal{A}$ such that

$$j[xy] = j(x)j(y) - j(y)j(x) \quad \text{all } x, y \in \mathfrak{g},$$

there is a unique homomorphism of algebras $f : \mathcal{U}(\mathfrak{g}) \to \mathcal{A}$ with $f \circ i = j$.

It is readily seen that any representation of $\mathfrak{g}$ gives a representation of $\mathcal{U}(\mathfrak{g})$, and conversely. In fact the map $i : \mathfrak{g} \to \mathcal{U}(\mathfrak{g})$ is injective and so $\mathfrak{g}$ can be regarded as a subspace of $\mathcal{U}(\mathfrak{g})$. A key property of $\mathcal{U}(\mathfrak{g})$ is the *Poincaré–Birkhoff–Witt basis theorem.* This asserts that if $x_1, \ldots, x_n$ is a basis of $\mathfrak{g}$ then the set of elements

$$x_1^{m_1} \cdots x_n^{m_n} \qquad m_1, \ldots, m_n \in \mathbb{Z}, \qquad m_i \geqslant 0$$

form a basis for $\mathcal{U}(\mathfrak{g})$. Each relation

$$[x_i x_j] = \sum_k \lambda_{ijk} x_k$$

in $\mathfrak{g}$ gives rise to a corresponding relation

$$x_i x_j - x_j x_i = \sum_k \lambda_{ijk} x_k$$

in $\mathcal{U}(\mathfrak{g})$.

$\mathcal{U}(\mathfrak{g})$ is a Noetherian algebra with no zero divisors.

8 Hopf algebra structures

The universal enveloping algebra $\mathcal{U}(\mathfrak{g})$ has more structure than that of an associative algebra. It also has the structure of a coalgebra. This means that there is an algebra homomorphism

$$\Delta : \mathcal{U}(\mathfrak{g}) \to \mathcal{U}(\mathfrak{g}) \otimes \mathcal{U}(\mathfrak{g})$$

called *comultiplication*, which is uniquely determined by the property

$$\Delta(x) = x \otimes 1 + 1 \otimes x \qquad \text{for all } x \in \mathfrak{g}.$$

The map Δ is coassociative.

There is also an algebra homomorphism

$$\epsilon : \mathcal{U}(\mathfrak{g}) \to k$$

called the *counit* map, uniquely determined by the properties

$$\epsilon(1) = 1\,,\ \ \epsilon(x) = 0 \text{ for all } x \in \mathfrak{g}\,.$$

The algebra and coalgebra structures on $\mathcal{U}(\mathfrak{g})$ make it into a bialgebra.

There is also an antiisomorphism of algebras

$$S : \mathcal{U}(\mathfrak{g}) \to \mathcal{U}(\mathfrak{g})$$

called the *antipode*, uniquely determined by the property

$$S(x) = -x \quad \text{for all } x \in \mathfrak{g}\,.$$

If $\mu : \mathcal{U}(g) \otimes \mathcal{U}(\mathfrak{g}) \to \mathcal{U}(\mathfrak{g})$ is the multiplication map and $\eta : k \to \mathcal{U}(\mathfrak{g})$ is given by $\eta(\lambda) = \lambda 1$ then we have a compatibility condition

$$\mu \circ (S \otimes 1) \circ \Delta = \mu \circ (1 \otimes S) \circ \Delta = \eta \circ \epsilon\,.$$

The bialgebra structure, together with the antipode map satisfying this compatibility condition, makes $\mathcal{U}(\mathfrak{g})$ into a *Hopf algebra.*

There is also a Hopf algebra associated to any affine algebraic group G over k. This is useful as it permits an approach to the study of affine algebraic groups which is independent of their embedding in an affine space k^n.

Given any affine variety $V \subset k^n$ let $\mathfrak{I}(V)$ be the ideal in the polynomial ring $k[x_1, \dots, x_n]$ of all polynomial functions which vanish on V. The quotient algebra

$$k[V] = k[x_1, \dots, x_n]/\mathfrak{I}(V)$$

is called the *affine algebra* of V. It is a finitely generated k-algebra with no non-zero nilpotent elements. Any homomorphism

$$V \xrightarrow{\phi} V'$$

of affine varieties induces a homomorphism of k-algebras

$$k[V'] \xrightarrow{\phi^*} k[V]$$

given by $\phi^*(f) = f \circ \phi$.

Now let G be an affine algebraic group. Then the affine algebra $k[G]$ has additional structure, induced from the group structure on G. Given $f \in k[G]$ there exist g_i, $h_i \in k[G]$ such that

$$f(xy) = \sum_i g_i(x) h_i(y) \qquad \text{for all } x, y \in G\,.$$

Then the map $\Delta : k[G] \to k[G] \otimes k[G]$ given by

$$\Delta(f) = \sum_i g_i \otimes h_i$$

is a comultiplication which is coassociative.

There is also a counit map, i.e. an algebra homomorphism

$$\epsilon : k[G] \to k$$

given by $\epsilon(f) = f(1)$.

Then there is an antipode map, i.e. an antiisomorphism of algebras given by

$$S : k[G] \to k[G]$$

where $Sf(x) = f(x^{-1})$ for all $f \in k[G]$, $x \in G$. The consistency conditions relating μ, Δ, η, ϵ, S are satisfied, and so $k[G]$ becomes a Hopf algebra.

The affine algebraic group G can be recovered from its Hopf algebra $k[G]$. Each $g \in G$ determines a k-algebra homomorphism from $k[G]$ to k given by $f \to f(g)$. This gives a bijection

$$G \to \mathrm{Hom}_{\mathbf{k\text{-}alg}}(k[G], k)\,.$$

Moreover each homomorphism $\phi : G_1 \to G_2$ of affine algebraic groups gives rise to a homomorphism $\phi^* : k[G_2] \to k[G_1]$ of the corresponding Hopf algebras. Thus we can reformulate the theory of affine algebraic groups in terms of Hopf algebras in a way which does not involve any embedding of the group in affine space. In fact the category of affine algebraic groups over the algebraically closed field k is anti-equivalent to the category of finitely generated commutative Hopf algebras over k with no non-zero nilpotent elements.

References

Borel, A. (1991) 'Linear algebraic groups', *Graduate Texts in Mathematics* **126**, Springer.

Humphreys, J.E. (1975) 'Linear algebraic groups', *Graduate Texts in Mathematics* **21**, Springer.

Springer, T.A. (1981) 'Linear algebraic groups', *Progress in Mathematics* **9**, Birkhäuser.

Weyl groups, affine Weyl groups and reflection groups

Raphaël Rouquier

UMR 9994 du CNRS, Université Paris 7–Denis Diderot, Case 7012, 2 Place Jussieu, F–75251 Paris Cedex 05, France

1 Introduction

This paper is a survey of some of the basic results pertaining to reflection groups.

In §2, we start with the basic concepts and properties of Coxeter groups, such as the Exchange Lemma and in §4 we construct the geometric representation. Sections 3 and 5 are devoted to finite real reflection groups and finite Coxeter groups and §6 concerns Weyl groups, which are crystallographic reflection groups. Weyl groups give rise to affine Weyl groups, studied in §7. The Iwahori-Hecke algebra of a Coxeter group is introduced in §8, after a discussion on braid groups. Finite complex reflection groups are the subject of §9, where we describe the infinite families. Finally, we explain in §10 how the topology of the hyperplane complement allows us to define braid groups and Iwahori-Hecke algebras for finite complex reflection groups.

This paper is expository : most proofs are to be found in [Bki] or [Hu] for §2-8 and in [BrMaRo] for §9-10.

2 Coxeter groups

Let W be a group and S a set of (distinct) generators of W of order 2. For $s, s' \in S$, we denote by $m_{s,s'} \in \{1, 2, \dots\} \cup \{\infty\}$ the order of the product ss'.

Definition 2.1 *The pair* (W, S) *is a* Coxeter system *if* W *has a presentation by generators and relations given by the set of generators* S *and the relations :*

$$s^2 = 1 \text{ for } s \in S,$$

$$\underbrace{ss'ss'\cdots}_{m_{ss'}\text{ terms}} = \underbrace{s'ss's\cdots}_{m_{ss'}\text{ terms}} \text{ for those } s, s' \in S \text{ such that } m_{s,s'} \text{ is finite.}$$

We then say also that W is a Coxeter group.

The relations $ss'ss'\cdots = s'ss's\cdots$ are called *braid relations*.

The *rank* of the system is the cardinality of S.

The *matrix* of the Coxeter system (W, S) is $(m_{s,s'})_{s,s' \in S}$; it has values in $\{1, 2, \ldots\} \cup \{\infty\}$.

This is a symmetric matrix with diagonal entries 1 and off-diagonal entries at least 2. A matrix with such properties is called a *Coxeter matrix*. We will see (Theorem 4.1) that every Coxeter matrix is the matrix of a Coxeter system (in a group given by generators and relations as in the definition, with $(m_{s,s'})$ an abstract Coxeter matrix, it isn't obvious that ss' will have order $m_{s,s'}$).

The *graph* associated with (W, S) is the graph with set of vertices S and edges $\{s, s'\}$ when $m_{s,s'} \geq 3$. Furthermore, the edge is then labelled by $m_{s,s'}$.

Some examples.

(i) The symmetric group $\mathfrak{S}_n = \mathfrak{S}(\{1, 2, \ldots, n\})$. Let $s_i = (i, i+1)$ and $S_{\mathfrak{S}_n} = \{s_1, \ldots, s_{n-1}\}$. Then, $(\mathfrak{S}_n, S_{\mathfrak{S}_n})$ is a Coxeter system (of type A_{n-1}) with graph

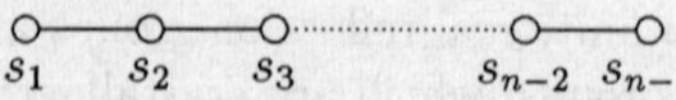

(the label is omitted when it is 3).

It has rank $n-1$.

(ii) The hyperoctahedral group B_n, *i.e.*, the group of $n \times n$ monomial matrices with non-zero entries in $\{\pm 1\}$. It contains $\mathfrak{S}_n$, viewed as the group of permutation matrices, as a subgroup. Let $s_0 = \mathrm{diag}(-1, 1, \ldots, 1)$ and $S_{B_n} = \{s_0, s_1, \ldots, s_{n-1}\}$, with s_i, $i \geq 1$ as in (i). Then, (B_n, S_{B_n}) is a Coxeter system of rank n with graph

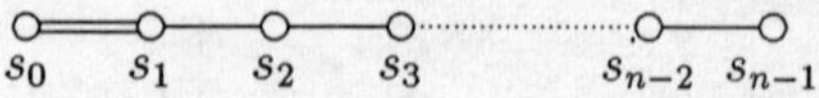

(where ○═○ stands for ○—4—○).

(iii) The dihedral group $I_2(m)$: this is the symmetry group of a regular m-gon (*i.e.*, the subgroup of the group of isometries of the plane fixing the m-gon), $m \geq 2$.

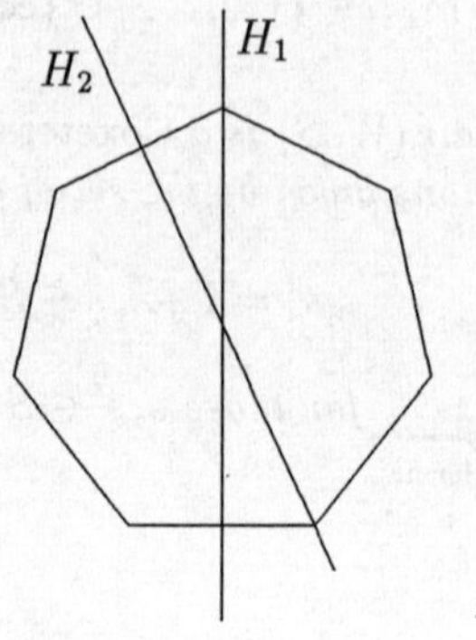

Let H_1 be a line containing the center of the polygon and one of its vertices. Let H_2 be a line containing the center of the polygon and such that the angle between H_1 and H_2 is π/m. $I_2(m)$ is generated by the orthogonal reflections t_1 and t_2 with respect to H_1 and H_2. $(I_2(m), \{t_1, t_2\})$ is a Coxeter system with graph

$$\underset{t_1}{\circ}\overset{m}{\text{---}}\underset{t_2}{\circ}\ (m \geq 3) \quad \text{or} \quad \underset{t_1}{\circ} \qquad \underset{t_2}{\circ}\ (m = 2).$$

The group $I_2(m)$ has a decomposition $I_2(m) = \langle t_1 t_2 \rangle \rtimes \langle t_1 \rangle$. The subgroup $\langle t_1 t_2 \rangle$ is the subgroup of rotations, it has order m. The action of $\langle t_1 \rangle \simeq \{\pm 1\}$ on $\mathbf{Z}/m\mathbf{Z}$ in this decomposition is given by multiplication.

This suggests a construction for $m = \infty$: we denote by $\tilde{A}_1$ the group $\mathbf{Z} \rtimes \{\pm 1\}$, where $\{\pm 1\}$ acts by multiplication on $\mathbf{Z}$. Let $t_1 = (0, -1)$ and $t_2 = (-1, -1)$. Then, $(\tilde{A}_1, \{t_1, t_2\})$ is a Coxeter system with graph

$$\underset{t_1}{\circ}\overset{\infty}{\text{---}}\underset{t_2}{\circ}$$

The dihedral groups are the groups $I_2(m)$, $2 \leq m < \infty$ and $\tilde{A}_1$.

Note that every rank 2 Coxeter system is isomorphic to the Coxeter system of a dihedral group. In particular, the Coxeter systems for $\mathfrak{S}_3$ and $I_2(3)$ are isomorphic, as well as those for B_2 and $I_2(4)$.

The following theorem [Bki, Chap. IV, §1, Théorème 2] is an easy consequence of Theorem 4.1 below :

Theorem 2.2 *Let (W, S) be a Coxeter system, S' a subset of S and W' the subgroup of W generated by S'. Then, (W', S') is a Coxeter system with Coxeter matrix the submatrix of the Coxeter matrix of (W, S) given by S'.*

A Coxeter system is *irreducible* if its associated graph is connected. All systems in the previous examples are irreducible, except $I_2(2)$.

If S is the disjoint union of two subsets S_1 and S_2 and no vertex of S_1 is connected to a vertex of S_2, then $W = W_1 \times W_2$, where W_i is the subgroup of W generated by S_i.

Remark 1 *Note that for m odd, $I_2(2m) \simeq I_2(m) \times \mathfrak{S}_2$, but $(I_2(2m), \{s, s'\})$ is nevertheless irreducible for $m > 1$!*

Let $w \in W$. The *length* of w, $l(w)$, is the smallest integer m such that w is the product of m elements of S.

A decomposition $w = s_1 \cdots s_m$ with $s_1, \ldots, s_m \in S$ is *reduced* if $m = l(w)$.

Theorem 2.3 *Let $w = s_1 \cdots s_m$ with $s_1, \ldots, s_m \in S$. Then, there is a subset $I = \{i_1 < i_2 < \cdots < i_k\}$ of $\{1, \ldots, m\}$ with $k = l(w)$ elements such that $w = s_{i_1} \cdots s_{i_k}$.*

This theorem is a direct consequence of the exchange lemma [Bki, Chap. IV, §1, Proposition 4] :

Lemma 2.4 *Let $w = s_1 \cdots s_m$ be a reduced decomposition ($s_1, \ldots, s_m \in S$). Let $s \in S$. Then, one of the following assertion holds :*

(i) $l(sw) = l(w) + 1$ and $ss_1 \cdots s_m$ is a reduced decomposition of sw

(ii) $l(sw) = l(w) - 1$ and there exists $j \in \{1, \ldots, m\}$ such that $s_1 \cdots s_{j-1}s_{j+1} \cdots s_m$ is a reduced decomposition of sw and $ss_1 \cdots s_{j-1}s_{j+1} \cdots s_m$ is a reduced decomposition of w.

This lemma actually characterizes the Coxeter systems amongst the pairs (W, S), where S is a set of generators of order 2 of a group W [Bki, Chap. IV, §1, Théorème 1].

3 Real reflection groups

Let V be a finite dimensional real vector space. A reflection of V is an automorphism of order 2 whose set of fixed points is a hyperplane. A finite reflection group W in V is a finite subgroup of $GL(V)$ generated by reflections.

The group W is *crystallographic* if there is a W-invariant $\mathbf{Z}$-lattice of V, *i.e.*, if there exists a free $\mathbf{Z}$-submodule L of V stable under W such that the canonical map $L \otimes_{\mathbf{Z}} \mathbf{R} \to V$ is an isomorphism.

Note that this amounts to the existence of a W-stable $\mathbf{Q}$-structure on V, *i.e.*, a $\mathbf{Q}$-subspace $V_{\mathbf{Q}}$ of V stable under W such that the canonical map $V_{\mathbf{Q}} \otimes_{\mathbf{Q}} \mathbf{R} \to V$ is an isomorphism [Bki, Chap. VI, §2, Proposition 9].

Let $\mathcal{A}$ be the set of reflecting hyperplanes of W — *i.e.*, the set of $\ker(s-1)$, where s is a reflection of W.

Then, $V - \bigcup_{H \in \mathcal{A}} H$ is in general non-connected : its connected components are the *chambers* of W.

Theorem 3.1 ([Bki, Chap. V, §3, Théorèmes 1 et 2]) *The group W acts simply transitively on the set of chambers ; the closure of a chamber is a fundamental domain for the action of W on V.*

Let C_1 be a chamber and S the set of reflections with respect to the walls of C_1 (a wall of C_1 is a hyperplane in $\mathcal{A}$ whose intersection with the closure of C_1 has codimension 1 in V).

Theorem 3.2 ([Bki, Chap. V, §3, Théorème 1]) *The pair (W, S) is a Coxeter system.*

Taking into account the choice of the chamber C_1, the chambers are now parametrized by W. The chamber C_w corresponding to $w \in W$ is $w(C_1)$.

A *gallery* of length n is a sequence $D_0, \ldots, D_n$ of adjacent chambers (*i.e.*, the intersection of the closures of D_i and D_{i+1} has codimension 1 in V). The following result can be deduced from [Bki, Chap. V, §3, Théorème 1] :

Proposition 3.3 *The minimal length of a gallery from C_0 to C_w is $l(w)$.*

Example : the chamber system for the group A_2.

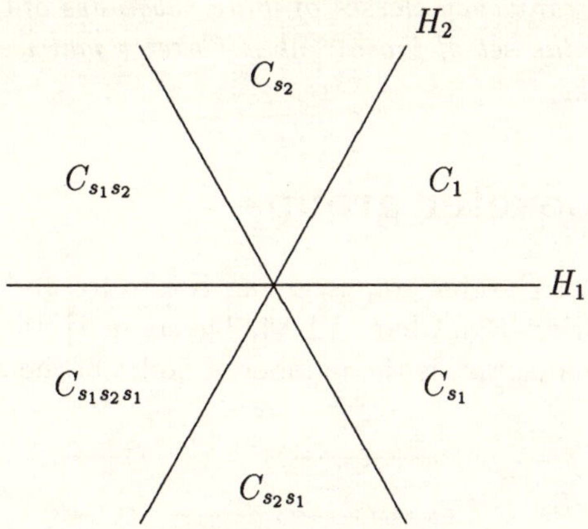

Here, s_i is the orthogonal reflection with respect to H_i. The group W generated by s_1 and s_2 is a Coxeter group of type A_2.

4 Coxeter groups as reflection groups

Let S be a set and $M = (m_{s,s'})_{s,s' \in S}$ a Coxeter matrix. Let $V = \mathbf{R}^S$ and denote by $\{e_s\}_{s \in S}$ its canonical basis.

Define a bilinear form B_M on V by

$$B_M(e_s, e_{s'}) = -\cos \frac{\pi}{m_{s,s'}}.$$

(Note that $B_M(e_s, e_s) = 1$).

Let ρ_s be the reflection in V given by

$$\rho_s(x) = x - 2B_M(e_s, x)e_s.$$

One has $V = \mathbf{R}e_s \oplus H_s$, where H_s is the hyperplane orthogonal to e_s.

Let W be the group with set of generators S and relations

$$s^2 = 1$$

$$\underbrace{ss'ss'\cdots}_{m_{ss'}\ terms} = \underbrace{s'ss's\cdots}_{m_{ss'}\ terms} \text{ for those } s, s' \in S \text{ such that } m_{s,s'} \neq \infty.$$

Theorem 4.1 ([Bki, Chap. V, §4.3 et §4.4]) *The map $s \mapsto \rho_s$ extends to an injective group morphism $W \to GL(V)$, the reflection representation of W. Furthermore, (W, S) is a Coxeter system.*

When S is finite, B_M is positive definite if and only if W is finite. Summarizing Theorems 3.2 and 4.1, we deduce

Theorem 4.2 *The constructions of §3 and §4 give rise to inverse bijections between the set of conjugacy classes of finite subgroups of $GL_n(\mathbf{R})$ generated by reflections and the set of those rank n Coxeter matrices giving rise to a finite Coxeter group.*

5 Finite Coxeter groups

The classification of Coxeter graphs giving rise to irreducible finite Coxeter groups is the following [Bki, Chap. VI, §4, Théorème 1] (the number attached to the name of the diagram is the number of nodes of the diagram) :

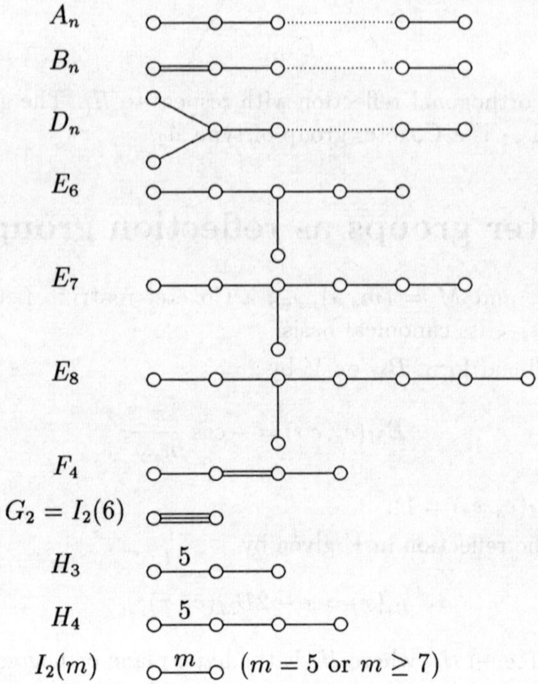

In the list above, all the groups are crystallographic except H_3, H_4 and $I_2(m)$, $m = 5$ or $m \geq 7$.

6 Root systems and Weyl groups

Let V be a finite dimensional real vector space, Φ a finite subset of V and $\Phi^\vee$ a finite subset of V^* parametrized by $\Phi : \Phi \to \Phi^\vee$, $\alpha \mapsto \alpha^\vee$.

Assume

(1) the vector space V is generated by Φ

(2) for all $\alpha \in \Phi$, we have $\langle \alpha^\vee, \alpha \rangle = 2$ and the reflection $s_\alpha : V \to V$, $x \mapsto x - \langle \alpha^\vee, x \rangle \alpha$ stabilizes Φ

(3) we have $\alpha^\vee(\Phi) \subset \mathbf{Z}$ for all $\alpha \in \Phi$

(4) for $\alpha \in \Phi$, we have $2\alpha \notin \Phi$.

Then, Φ is a *root system* in V (sometimes called reduced, because of (4)). Note that given Φ, there is at most one set $\Phi^\vee$ parametrized by Φ with the required properties.

If $\Phi = \Phi_1 \cup \Phi_2$ and Φ_i (together with $\Phi_i^\vee = \{\alpha^\vee\}_{\alpha \in \Phi_i}$) is a root system in V_i, the subspace of V generated by Φ_i, for $i \in \{1, 2\}$, then we say that Φ is the direct sum of the root systems Φ_1 and Φ_2. The root system Φ is *irreducible* if it is non-empty and it is not the direct sum of two non-empty root systems.

The *Weyl group* of the root system Φ is the subgroup of $GL(V)$ generated by the reflections s_α for $\alpha \in \Phi$. Note that W is a crystallographic finite reflection group with **Z**-lattice the **Z**-submodule of V generated by Φ. A converse actually holds [Bki, Ch. VI, §2, Proposition 9] :

Proposition 6.1 *Let W be a crystallographic reflection group in a finite dimensional real vector space V. Then, there is a root system Φ in V with Weyl group W.*

Note that if W is irreducible, then the root system Φ is unique up to isomorphism if and only if W is not of type B_n, $n \geq 3$ (cf Remark 2).

Let C be a chamber of W with walls $L_1, \ldots, L_n$. Then, there is a unique root $\alpha_i \in \Phi$ orthogonal to L_i and lying in the same half-space determined by L_i as C.

The set $\Delta = \{\alpha_i\}_{1 \leq i \leq n}$ is called a *basis* of Φ.

Let $\Phi^+ = \{\alpha \in \Phi | \alpha = \sum n_i \alpha_i, n_i \geq 0\}$ (the *positive* roots) and $\Phi^- = \{\alpha \in \Phi | \alpha = \sum n_i \alpha_i, n_i \leq 0\}$ (the *negative* roots).

Proposition 6.2 ([Bki, Chap. VI, §1, Théorèmes 2 et 3]) *The set* Δ *is a basis of* V *and* $\Phi = \Phi^+ \cup \Phi^-$.

The *Cartan matrix* is $(\langle \alpha, \beta^\vee \rangle)_{\alpha,\beta\in\Delta}$.
Define $S = \{s_\alpha\}_{\alpha\in\Delta}$. Then, (W, S) is a Coxeter system.

The rank 2 root systems

Type $A_1 \times A_1$

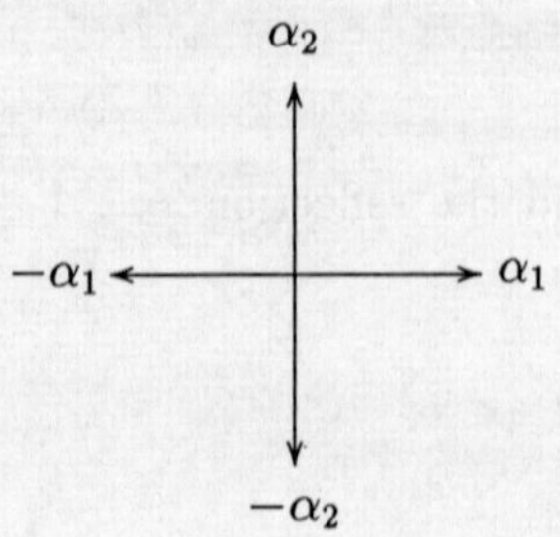

Type A_2

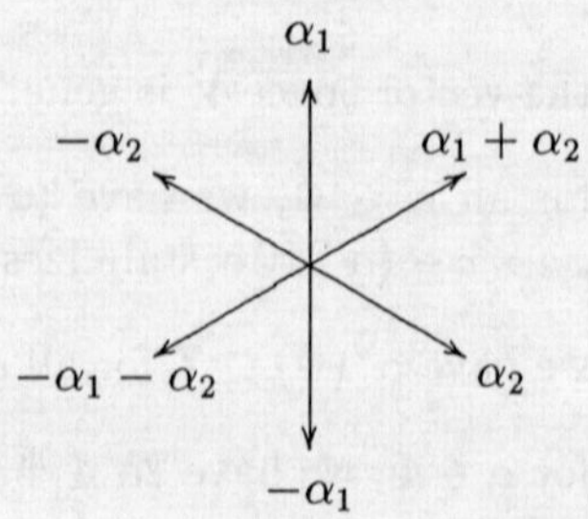

Type $B_2 = C_2$

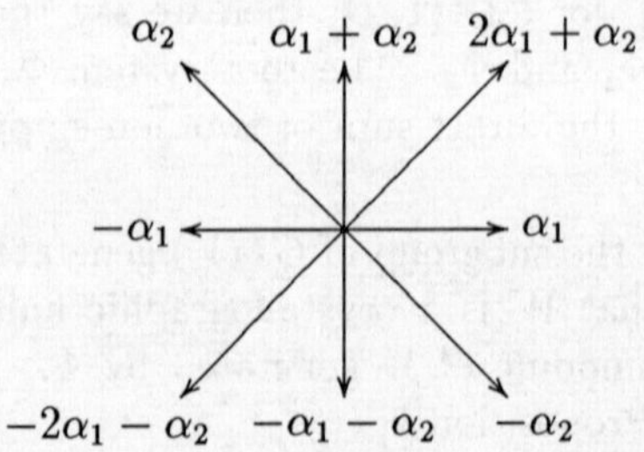

Type G_2

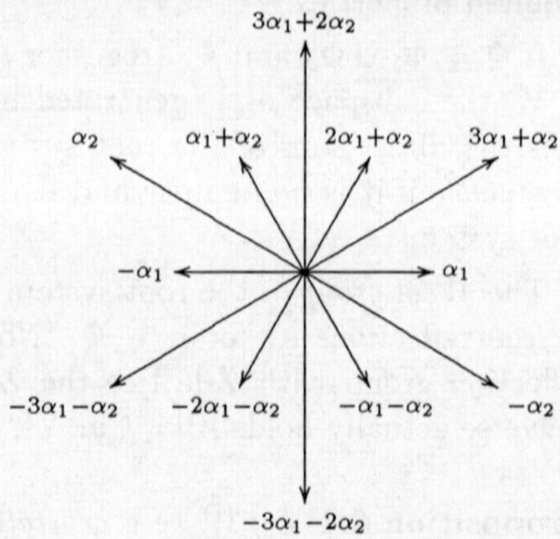

The action of W on Φ gives another interpretation of the length function [Bki, Chap. VI, §1, Corollaire 2 de la Proposition 17] :

Proposition 6.3 *Let* $w \in W$. *The cardinality of* $\Phi^- \cap w(\Phi^+)$ *is the length* $l(w)$.

The set $\Phi^\vee$ defines a root system in V^* (the root system *inverse* or *dual* to Φ). There is an isomorphism of groups

$$W(\Phi) \to W(\Phi^\vee),$$
$$u \mapsto {}^t u^{-1}$$

sending s_α on $s_{\alpha^\vee}$. Through this isomorphism, $W(\Phi)$ operates on V^*.

Remark 2 *Note that the root systems (V, Φ) and (V^*, Φ^*) are not isomorphic in general : for example, the root system of type C_n is the inverse of the root system of type B_n ; when $n \geq 3$, these root systems are not isomorphic.*

7 Affine Weyl groups

Let Φ be a root system in a finite dimensional real vector space V. We construct a subgroup of the group $\mathrm{Aff}(V^*)$ of affine transformations of V^* as follows :

For $\alpha \in \Phi$ and $k \in \mathbf{Z}$, let $H_{\alpha,k}$ be the affine hyperplane of V^* defined by

$$H_{\alpha,k} = \{x \in V^* | \langle \alpha, x \rangle = k\}.$$

Let $s_{\alpha,k}$ be the orthogonal reflection with respect to $H_{\alpha,k}$:

$$s_{\alpha,k}(x) = x - (\langle \alpha, x \rangle - k)\alpha^\vee.$$

The affine Weyl group associated to Φ is the subgroup $\mathcal{W}$ of $\mathrm{Aff}(V^*)$ generated by the $s_{\alpha,k}$, $\alpha \in \Phi$, $k \in \mathbf{Z}$.

Let Q be the subgroup of $\mathrm{Aff}(V^*)$ generated by the translations by elements of $\Phi^\vee$.

Proposition 7.1 ([Bki, Chap. VI, §2, Proposition 1]) *We have $\mathcal{W} = Q \rtimes W$.*

An *alcove* is a connected component of $V^* - \bigcup_{\alpha\in\Phi,k\in\mathbf{Z}} H_{\alpha,k}$ (note that the set $\{H_{\alpha,k}\}_{\alpha\in\Phi,k\in\mathbf{Z}}$ is the set of reflecting hyperplanes of $\mathcal{W}$).

Theorem 7.2 ([Bki, Chap. VI, §2.1]) *The group $\mathcal{W}$ acts simply transitively on the set of alcoves. The closure of an alcove is a fundamental domain for the action of $\mathcal{W}$ on V^*.*

Let C be a chamber for W. Then, there is a unique alcove $A \subset C$ such that 0 is in the closure of A.

Let $\tilde{S}$ be the set of reflections with respect to the walls of A. The pair $(\mathcal{W}, \tilde{S})$ is Coxeter group.

If Φ is irreducible, there is a root $\tilde{\alpha} = \sum n_i \alpha_i$ in Φ such that if $\beta \in \Phi$, $\beta = \sum m_i \alpha_i$, then $m_i \leq n_i$: $\tilde{\alpha}$ is the *highest root*. This root is orthogonal to the wall of A which doesn't contain 0 [Bki, Chap. VI, §1, Proposition 25].

The length of an element $w \in \mathcal{W}$ (relatively to $\tilde{S}$) is the minimal length of a gallery of alcoves from A to $w(A)$.

Example : Type $\tilde{A}_2$.

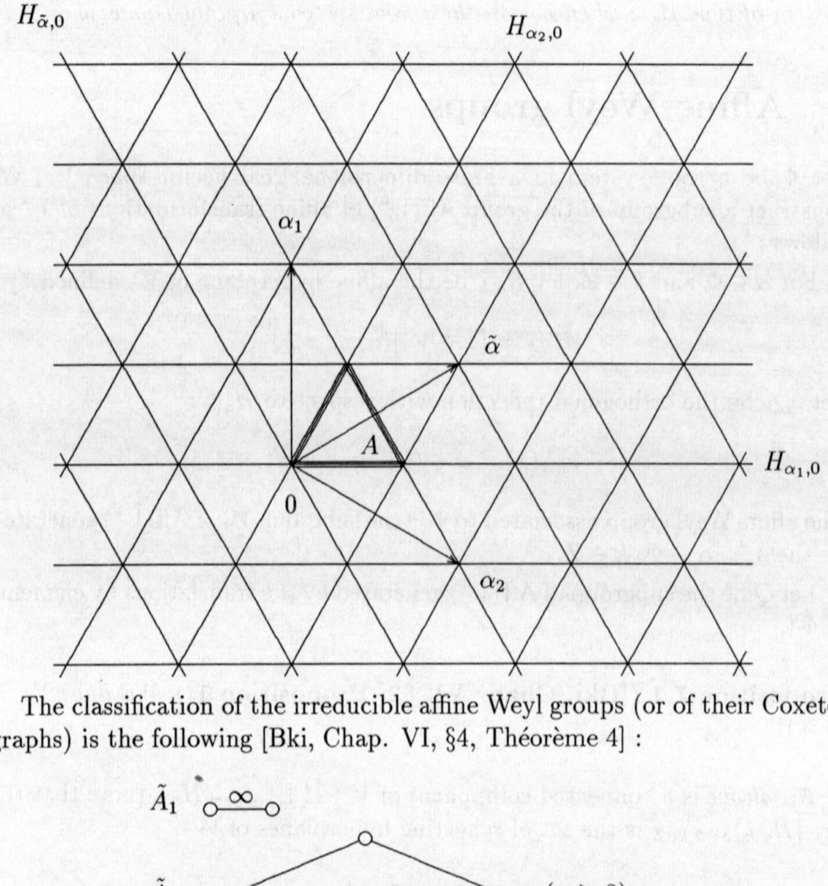

The classification of the irreducible affine Weyl groups (or of their Coxeter graphs) is the following [Bki, Chap. VI, §4, Théorème 4] :

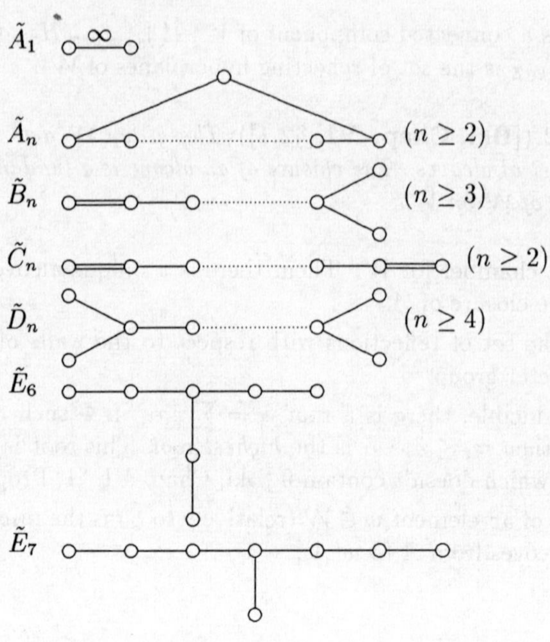

$\tilde{E}_8$ 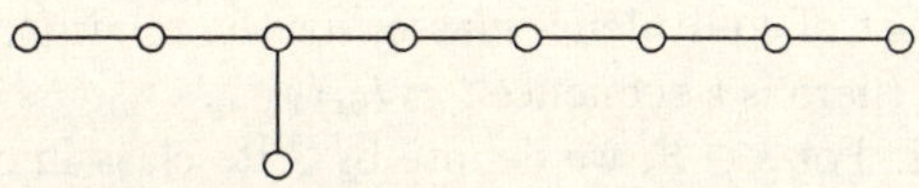

$\tilde{F}_4$ 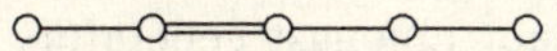

$\tilde{G}_2$ 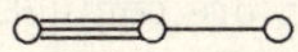

8 Braid groups and Iwahori-Hecke algebras

Let (W, S) be a Coxeter system.

Definition 8.1 *The* braid monoid B_W^+ *associated with* (W, S) *is the monoid with set of generators* $\{\sigma_s^+\}_{s\in S}$ *and relations :*

$$\underbrace{\sigma_s^+\sigma_{s'}^+\sigma_s^+\sigma_{s'}^+\cdots}_{m_{ss'}\ terms} = \underbrace{\sigma_{s'}^+\sigma_s^+\sigma_{s'}^+\sigma_s^+\cdots}_{m_{ss'}\ terms} \quad \textit{for } s, s' \in S \textit{ such that } m_{s,s'} \textit{ is finite.}$$

The braid group B_W *associated with* (W, S) *is the group with set of generators* $\{\sigma_s\}_{s\in S}$ *and relations :*

$$\underbrace{\sigma_s\sigma_{s'}\sigma_s\sigma_{s'}\cdots}_{m_{ss'}\ terms} = \underbrace{\sigma_{s'}\sigma_s\sigma_{s'}\sigma_s\cdots}_{m_{ss'}\ terms} \quad \textit{for } s, s' \in S \textit{ such that } m_{s,s'} \textit{ is finite.}$$

The following result is due to Deligne [De, Proposition 4.17] and Brieskorn-Saito [BrSa, Proposition 5.5] :

Theorem 8.2 *Assume W is finite. Then, the morphism $B_W^+ \to B_W$ given by $\sigma_s^+ \mapsto \sigma_s$ is injective.*

Thanks to this result, we can identify B_W^+ with the submonoid of B_W consisting of those elements which can be written as products of generators σ_s, when W is finite.

Let $p : B_W^+ \to W$ be given by $\sigma_s \mapsto s$.

The exchange lemma (Lemma 2.4) has the following consequence [Bki, Chapitre IV, §1, Proposition 5] :

Proposition 8.3 *Let $w = s_1 \cdots s_m = s'_1 \cdots s'_m$ be two reduced expressions of $w \in W$. Then, $\sigma_{s_1}^+ \cdots \sigma_{s_m}^+ = \sigma_{s'_1}^+ \cdots \sigma_{s'_m}^+$.*

This allows the construction of a very nice section q of p (q is not multiplicative !) : given $w \in W$ and $w = s_1 \cdots s_m$ a reduced expression of w, we put $q(w) = \sigma_{s_1}^+ \cdots \sigma_{s_m}^+$. Thanks to the last proposition, this is independent of the choice of the reduced expression of w. We have $pq(w) = w$.

Let $\bar{S}$ be the set of equivalence classes for the relation defined by : $s, t \in S$ are equivalent if there is a sequence $t = t_0, t_1, \ldots, t_m = s$ such that $m(t_i, t_{i+1})$ is finite and odd. For $s \in S$, we denote by $\bar{s}$ its class in $\bar{S}$.

One can then read off conjugacy amongst generators of W and B_W from the Coxeter diagram. Two elements $s, t \in S$ are conjugate in W if and only if $\bar{s} = \bar{t}$. Similarly, σ_s and σ_t are conjugate in B_W if and only if $\bar{s} = \bar{t}$. So, we have :

Proposition 8.4 *The map $\sigma_s \mapsto 1_{\bar{s}}$ extends to a group morphism $B_W \to \mathbf{Z}^{\bar{S}}$. Its kernel is the derived subgroup of B_W,* i.e., *this morphism identifies the largest abelian quotient B_W^{ab} of B_W with $\mathbf{Z}^{\bar{S}}$. Similarly, W^{ab} is isomorphic to $(\mathbf{Z}/2\mathbf{Z})^{\bar{S}}$.*

Examples (using the notations from the examples of §2).

- $\mathfrak{S}_n : \bar{S} = \{\bar{s}_1\}$.
- $B_n : \bar{S} = \{\bar{s}_0, \bar{s}_{n-1}\}$ $(n \geq 2)$.
- $I_2(m) : \bar{S} = \{\bar{t}_1\}$ for odd m, $\bar{S} = \{\bar{t}_1, \bar{t}_2\}$ for even m.

Let $\{q_{\bar{s}}\}_{\bar{s} \in \bar{S}}$ and $\{q'_{\bar{s}}\}_{\bar{s} \in \bar{S}}$ be two sets of indeterminates. Let $\tilde{\mathcal{O}} = \mathbf{Z}[q_{\bar{s}}, q'_{\bar{s}}]_{\bar{s} \in \bar{S}}$.

The Iwahori-Hecke algebra $\tilde{\mathcal{H}} = \tilde{\mathcal{H}}(W)$ of (W, S) is the $\tilde{\mathcal{O}}$-algebra with generators T_s, $s \in S$, and relations

$$(T_s - q_{\bar{s}})(T_s - q'_{\bar{s}}) = 0 \text{ for } s \in S$$

$$\underbrace{T_s T_{s'} T_s T_{s'} \cdots}_{m_{ss'} \text{ terms}} = \underbrace{T_{s'} T_s T_{s'} T_s \cdots}_{m_{ss'} \text{ terms}} \text{ for } s, s' \in S \text{ such that } m_{s,s'} \text{ is finite.}$$

This is the quotient of the monoid algebra $\tilde{\mathcal{O}} B_W^+$ by the ideal generated by the elements $(T_s - q_{\bar{s}})(T_s - q'_{\bar{s}})$ for $s \in S$. Let T_w be the image of σ_w^+ for $w \in W$.

The next theorem shows that $\tilde{\mathcal{H}}$ is a deformation of $\mathbf{Z}W$ [Bki, Chap. IV, §2, Exercice 23] :

Theorem 8.5 *The algebra $\tilde{\mathcal{H}}$ is free over $\tilde{\mathcal{O}}$, with basis $\{T_w\}_{w \in W}$. The morphism $\tilde{\mathcal{H}} \otimes_{\tilde{\mathcal{O}}} \tilde{\mathcal{O}}/(q_{\bar{s}} - 1, q'_{\bar{s}} + 1)_{\bar{s} \in \bar{S}} \to \mathbf{Z}W$, $T_w \otimes 1 \mapsto w$, is an isomorphism.*

We assume from now on that W is finite.

Let S' be a subset of S and W' be the subgroup of W generated by S'. Then, by [De],

- the submonoid of B_W^+ generated by $\{\sigma_s^+\}_{s \in S'}$ is isomorphic to $B_{W'}^+$,

- the subgroup of B_W generated by $\{\sigma_s\}_{s\in S'}$ is isomorphic to $B_{W'}$,
- the specialization of the subalgebra of $\tilde{\mathcal{H}}(W)$ generated by $\{T_s\}_{s\in S'}$ obtained by sending to 0 those parameters not associated to elements $\bar{s}, s \in S'$, is isomorphic to the specialization of $\tilde{\mathcal{H}}(W')$ given by identifying those parameters associated to elements of $\bar{S}'$ which become equal in $\bar{S}$.

In several applications, the Iwahori-Hecke algebra arises with invertible parameters. Then, without loss of generality, one may assume one of the two parameters $q_{\bar{s}}, q'_{\bar{s}}$ to be -1. So, let $\mathcal{O} = \tilde{\mathcal{O}}[q_{\bar{s}}^{-1}]_{\bar{s}\in\bar{S}}/(q'_{\bar{s}}+1)_{\bar{s}\in\bar{S}} \simeq \mathbf{Z}[q_{\bar{s}}, q_{\bar{s}}^{-1}]_{\bar{s}\in\bar{S}}$ and $\mathcal{H} = \tilde{\mathcal{H}} \otimes_{\tilde{\mathcal{O}}} \mathcal{O}$.

The Iwahori-Hecke algebra has a trace map $\tau : \mathcal{H} \to \mathcal{O}$ given by $\tau(T_w) = \delta_{1,w}$ (*i.e.*, we have $\tau(hh') = \tau(h'h)$ for $h, h' \in \mathcal{H}$).

Denote by "ind" the one-dimensional representation $\mathcal{H} \to \mathcal{O}$ given by $\mathrm{ind}(T_s) = q_{\bar{s}}$.

Proposition 8.6 *Given $w, w' \in W$, one has $\tau(T_w T_{w'}) = \delta_{w^{-1},w'}\mathrm{ind}(T_w)$.*

This means that the set $\{\mathrm{ind}(T_w)^{-1}T_{w^{-1}}\}_{w\in W}$ is the dual basis of $\{T_w\}_{w\in W}$ with respect to τ.

More conceptually, the trace τ gives a structure of symmetric algebra to $\mathcal{H}$, *i.e.*, the morphism :

$$\begin{aligned} \mathcal{H} &\to \mathrm{Hom}_{\mathcal{O}}(\mathcal{H}, \mathcal{O}) \\ h &\mapsto (h' \mapsto \tau(hh')) \end{aligned}$$

is an isomorphism.

Together with the fact that $\mathcal{H}$ is a deformation of $\mathbf{Z}W$, this explains the structure of $\mathcal{H}$ over an algebraic closure K of the field of fractions of $\mathcal{O}$ (Tits' deformation theorem) [Bki, Chap. IV, §2, Exercice 27] :

Theorem 8.7 *The algebra $\mathcal{H} \otimes_{\mathcal{O}} K$ is semi-simple and isomorphic to KW.*

Much more precise is the following rationality theorem (Benard, Springer, Benson-Curtis, Hoefsmit, Lusztig..., cf [Ge]) :

Theorem 8.8 *Assume W is a finite Weyl group. Then, the algebra $\mathbf{Q}W$ is isomorphic to a direct product of matrix algebras over $\mathbf{Q}$ and the algebra $\mathcal{H} \otimes_{\mathcal{O}} \mathbf{Q}(\sqrt{q_{\bar{s}}})_{\bar{s}\in\bar{S}}$ is isomorphic to a direct product of matrix algebras over $\mathbf{Q}(\sqrt{q_{\bar{s}}})_{\bar{s}\in\bar{S}}$.*

The theorem above generalizes to finite Coxeter groups : if W is a finite reflection group over $K \subset \mathbf{R}$, then KW is isomorphic to a product of matrix algebras over K and $\mathcal{H} \otimes_{\mathcal{O}} K(\sqrt{q_{\bar{s}}})_{\bar{s}\in\bar{S}}$ is isomorphic to a direct product of matrix algebras over $K(\sqrt{q_{\bar{s}}})_{\bar{s}\in\bar{S}}$.

9 Pseudo-reflection groups

Let V be a finite dimensional vector space over a characteristic zero field K. A *pseudo-reflection* of V is an automorphism of finite order whose set of fixed points is a hyperplane.

Let G be a finite subgroup of $GL(V)$.

We denote by $S(V)$ the symmetric algebra of V.

The following theorem is due to Shephard-Todd and Chevalley. It shows that the regularity of the ring of invariants $S(V)^G$ characterizes pseudo-reflection groups [Bens, Theorem 7.2.1] :

Theorem 9.1 *The following assertions are equivalent :*

(i) The algebra $S(V)^G$ is a polynomial algebra.

(ii) The group G is generated by pseudo-reflections.

(iii) The $S(V)^G[G]$-module $S(V)$ is free of rank one.

When $K \subset \mathbf{R}$, a pseudo-reflection is actually a reflection. When $K = \mathbf{C}$, a pseudo-reflection need not have order 2 ; a group generated by complex pseudo-reflections is then called a *complex reflection group*.

The rationality theorem 8.8 for representations of Weyl groups or finite Coxeter groups extends to pseudo-reflection groups : the group algebra of G over K is a direct product of matrix algebras over K [Bena, Bes].

The irreducible complex reflection groups have been classified by Shephard and Todd [ShTo]. There are two infinite series : the groups $A_n \simeq \mathfrak{S}_{n+1}$, the groups $G(p, q, n)$ and 34 exceptional groups (the dimension of an exceptional group is at most 8).

Let us describe the groups $G(p, q, n)$ ($p > 1$, $q \geq 1$, $n \geq 1$ and $q|p$). It turns out that these groups have nice presentations, generalizing in some sense the presentation of Coxeter groups and sharing some of their properties. In particular, these groups have a presentation given by a set S consisting of n or $n + 1$ pseudo-reflections and two kinds of relations :

- braid relations (homogeneous relations)
- finite order relations.

The group given by the same presentation, but without the finite order relations can be seen as an analog of the braid group defined in §8 for real reflection groups. We will come back to this in §10.

9.1 $G(p,1,n)$

First, $G(p,1,n)$ is the group of n by n monomial complex matrices whose non-zero entries are p-th roots of unity. This group has a semi-direct product decomposition $G(p,1,n) = (\mathbf{Z}/p\mathbf{Z})^n \rtimes \mathfrak{S}_n \simeq (\mathbf{Z}/p\mathbf{Z}) \wr \mathfrak{S}_n$, where $\mathfrak{S}_n$ is the subgroup of permutation matrices and $(\mathbf{Z}/p\mathbf{Z})^n$ is the subgroup of diagonal matrices.

Let $s_0 = \mathrm{diag}(\zeta, 1, \ldots, 1)$, where ζ is a primitive p-th root of unity. Keeping the notations of §2, Example (i), one sees that $G(p,1,n)$ is generated by the set of pseudo-reflections $\{s_0, s_1, \ldots, s_{n-1}\}$. They satisfy the following relations :

$$\text{braid relations} \begin{cases} s_0 s_1 s_0 s_1 = s_1 s_0 s_1 s_0 & \\ s_i s_j = s_j s_i & \text{if } |i-j| > 1 \\ s_i s_{i+1} s_i = s_{i+1} s_i s_{i+1} & \text{for } i \geq 1 \end{cases}$$

$$\text{finite order relations} \begin{cases} s_0^p = 1 \\ s_i^2 = 1 \text{ for } i \geq 1. \end{cases}$$

Actually, this gives a presentation for $G(p,1,n)$ by generators and relations. A convenient way to encode the relations is to use a generalization of the Coxeter diagrams :

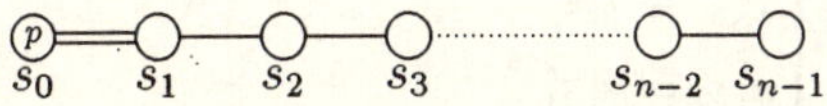

Note that $G(2,1,n) = B_n$ and the presentation above is the Coxeter presentation.

Now, for $q|p$, we define $G(p,q,n)$ as the subgroup of $G(p,1,n)$ consisting of matrices where the product of the non-zero entries is a (p/q)-th root of unity.

9.2 $G(p,p,n)$

Let us now look at $G(p,p,n)$. It is generated by the set of pseudo-reflections $\{s_1', s_1, \ldots, s_{n-1}\}$ where $s_1' = s_0 s_1 s_0^{-1}$. They satisfy the following relations :

$$\text{braid relations} \begin{cases} s_i s_j = s_j s_i & \text{if } |i-j| > 1 \\ s_1' s_i = s_i s_1' & \text{for } i \geq 3 \\ s_i s_{i+1} s_i = s_{i+1} s_i s_{i+1} & \text{for } i \geq 1 \\ s_1' s_2 s_1' = s_2 s_1' s_2 & \\ s_2 s_1' s_1 s_2 s_1' s_1 = s_1' s_1 s_2 s_1' s_1 s_2 & \\ \underbrace{s_1 s_1' s_1 s_1' \cdots}_{p \text{ terms}} = \underbrace{s_1' s_1 s_1' s_1 \cdots}_{p \text{ terms}} & \end{cases}$$

$$\text{finite order relations} \begin{cases} {s'_1}^2 = 1 \\ s_i^2 = 1 \text{ for } i \geq 1. \end{cases}$$

This gives a presentation of $G(p,p,n)$ by generators and relations. The relations may be encoded in the following diagram :

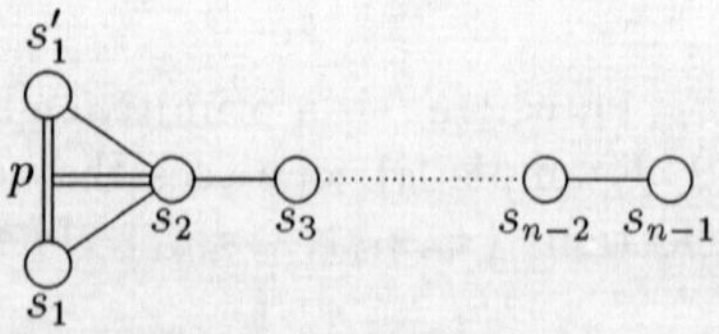

Note that $G(p,p,2) = I_2(p)$ and the presentation above is a Coxeter presentation. Also, $G(2,2,n) = D_n$ and the presentation above is a Coxeter presentation.

9.3 $G(p,q,n)$

Finally, let us consider $G(p,q,n)$ for $q|p$, $q \neq p$ and $q \neq 1$. We put $d = p/q$. This group is generated by the set of pseudo-reflections $\{s'_0, s'_1, s_1, \ldots, s_{n-1}\}$ where $s'_0 = s_0^q$. They satisfy the following relations :

$$\text{braid relations} \begin{cases} s_i s_j = s_j s_i & \text{if } |i-j| > 1 \\ s'_0 s_i = s_i s'_0 & \text{for } i \geq 2 \\ s'_1 s_i = s_i s'_1 & \text{for } i \geq 3 \\ s_i s_{i+1} s_i = s_{i+1} s_i s_{i+1} & \text{for } i \geq 1 \\ s'_1 s_2 s'_1 = s_2 s'_1 s_2 & \\ s'_0 s'_1 s_1 = s'_1 s_1 s'_0 & \\ s_2 s'_1 s_1 s_2 s'_1 s_1 = s'_1 s_1 s_2 s'_1 s_1 s_2 & \\ \underbrace{s_1 s'_0 s'_1 s_1 s'_1 s_1 \cdots}_{q+1 \text{ terms}} = \underbrace{s'_0 s'_1 s_1 s'_1 s_1 s'_1 \cdots}_{q+1 \text{ terms}} & \end{cases}$$

$$\text{finite order relations} \begin{cases} {s'_0}^d = 1 \\ {s'_1}^2 = 1 \\ s_i^2 = 1 \text{ for } i \geq 1. \end{cases}$$

We have obtained a presentation of $G(p,q,n)$ by generators and relations which we encode in the following diagram :

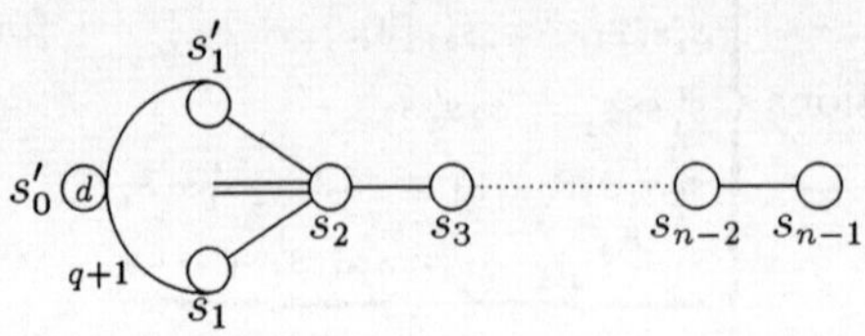

10 Topological construction of braid groups and Iwahori-Hecke algebras

Let V be a finite dimensional complex vector space and G a finite subgroup of $GL(V)$ generated by pseudo-reflections. Let $\mathcal{A}$ be the set of reflecting hyperplanes of G and $X = V - \bigcup_{H\in\mathcal{A}} H$. Let $\rho : X \to X/G$ be the projection map.

The following result is due to Steinberg [St] :

Theorem 10.1 *The group G acts freely on X,* i.e., *ρ is an unramified Galois covering.*

Let $x_0 \in X$. The *braid group* associated to G is $\mathcal{B}_G = \Pi_1(X/G, \rho(x_0))$ and the *pure braid group* associated to G is $\mathcal{P}_G = \Pi_1(X, x_0)$. Then, thanks to Steinberg's theorem, we have an exact sequence :

$$0 \to \mathcal{P}_G \xrightarrow{\rho_*} \mathcal{B}_G \to G \to 1.$$

10.1 The real case

Assume G is the complexification of a real reflection group, *i.e.*, there is a real vector space V' with $V = V' \otimes_{\mathbf{C}} \mathbf{R}$ and such that G is a subgroup of $GL(V')$. Let C_1 be a chamber of G (a connected component of $V' - \bigcup_{H\in\mathcal{A}} H \cap V'$) and take $x_0 \in C_1$. Let S be the set of reflections of G with respect to the walls of C_1. For $s \in S$, let γ_s be the path $[0,1] \to X$ defined by

$$\gamma_s(t) = \frac{x_0 + s(x_0)}{2} + \frac{x_0 - s(x_0)}{2} e^{i\pi t}.$$

Let τ_s be the class in $\mathcal{B}_G$ of $\rho(\gamma_s)$.

Brieskorn [Br] and Deligne [De] have proved the following theorem :

Theorem 10.2 *The map $\sigma_s \mapsto \tau_s$ induces an isomorphism $B_G \to \mathcal{B}_G$.*

10.2 The complex case

Let $H \in \mathcal{A}$. Let e_H be the order of the pointwise stabilizer of H in G. This is a cyclic group, generated by a pseudo-reflection s with non-trivial eigenvalue $\exp(2i\pi/e_H)$. Let $x_H \in X$. Let y_H be the intersection of H with the affine line containing x_H and $s(x_H)$. We assume x_H is close enough to H so that the closed ball with center y_H and radius $||x_H - y_H||$ does not intersect any H', $H' \in \mathcal{A}$, $H' \neq H$. Let α be a path from x_0 to x_H in X. Let λ be the path in X from x_H to $s(x_H)$ defined by

$$\lambda(t) = y_H + (x_H - y_H)e^{2i\pi t/e_H}.$$

We define the path γ from x_0 to $s(x_0)$ by

$$\gamma(t) = \begin{cases} \alpha(3t) & \text{for } 0 \leq t \leq 1/3 \\ \lambda(3t-1) & \text{for } 1/3 < t \leq 2/3 \\ s(\alpha(3-3t)) & \text{for } 2/3 < t \leq 1 \end{cases}$$

Let τ be the class of γ in $\mathcal{B}_G$: τ is called a *generator of the monodromy* associated to s (or to H). The image of τ in G is s.

Theorem 10.2 has a counterpart for complex reflection groups, based on a case by case study [BrMaRo] (six of the irreducible exceptional complex reflection groups are not covered by this approach).

Let us explain this for the group $G = G(p, q, n)$.

Theorem 10.3 *Assume $G = G(p, q, n)$. Then, for every $s \in S$, there is a generator of the monodromy τ_s associated to s, such that the group $\mathcal{B}_G$ has a presentation with set of generators $\{\tau_s\}_{s \in S}$ and relations the braid relations.*

10.3 Iwahori-Hecke algebras

An analogue of Proposition 8.4 for complex reflection groups is : $\mathcal{B}_G^{ab}$ is isomorphic to $\mathbf{Z}^{\mathcal{A}/G}$.

For $\mathcal{C} \in \mathcal{A}/G$ and $H \in \mathcal{C}$, we put $e_\mathcal{C} = e_H$. Let $\tilde{\mathcal{O}}$ be the polynomial ring over the integers on the set of variables $\{q_{\mathcal{C},j}\}_{\mathcal{C} \in \mathcal{A}/G, 0 \leq j \leq e_\mathcal{C}-1}$. We put $q_{H,j} = q_{\mathcal{C},j}$ for $H \in \mathcal{C}$ and $\mathcal{C} \in \mathcal{A}/G$.

For $H \in \mathcal{A}$, let τ_H be a generator of the monodromy associated to H. Let $\tilde{\mathcal{H}}$ be the quotient of the group algebra $\tilde{\mathcal{O}}\mathcal{B}_G$ by the ideal generated by the $(\tau_H - q_{H,0})(\tau_H - q_{H,1}) \cdots (\tau_H - q_{H,e_H-1})$, for $H \in \mathcal{A}$. Then, $\tilde{\mathcal{H}}$ is the *Iwahori-Hecke algebra* associated to G.

When G is the complexification of a real reflection group, then Theorem 10.2 induces an isomorphism with the Iwahori-Hecke algebra associated with the Coxeter system (G, S).

When $G = G(p, q, n)$, the algebra $\tilde{\mathcal{H}}$ is a deformation of the group algebra $\tilde{\mathcal{O}}G$ (it is free over $\tilde{\mathcal{O}}$, with rank $|G|$) [Ar].

References

[Ar] S. Ariki, *Representation theory of a Hecke algebra of $G(r, p, n)$*, J. of Alg. **106** (1994), 216–243.

[Bena] M. Benard, *Schur indices and splitting fields of the unitary reflection groups*, J. of Alg. **38** (1976), 318–342.

[Bens] D. Benson, "Polynomial invariants of finite groups", London Math. Soc. Lecture Note Series **190**, Cambridge University Press, 1993.

[Bes] D. Bessis, *Sur le corps de définition d'un groupe de réflexions complexe*, to appear in Comm. in Alg.

[Bki] N. Bourbaki, "Groupes et algèbres de Lie", Chapitres IV, V et VI, Hermann, Paris, 1968.

[Br] E. Brieskorn, *Die Fundamentalgruppe des Raumes der regulären Orbits einer endlichen komplexen Spiegelungsgruppe*, Inv. Math. **12** (1971), 57–61.

[BrSa] E. Brieskorn and K. Saito, *Artin-Gruppen und Coxeter-Gruppen*, Inv. Math. **17** (1972), 245–271.

[BrMaRo] M. Broué, G. Malle and R. Rouquier, *Complex reflection groups, braid groups, Hecke algebras*, preprint (1997).

[De] P. Deligne, *Les immeubles des groupes de tresses généralisés*, Inv. Math. **17** (1972), 273–302.

[Ge] M. Geck, *On the character values of Iwahori-Hecke algebras of exceptional type*, J. London Math. Soc. **68** (1994), 51–76.

[Hu] J.E. Humphreys, "Reflection groups and Coxeter groups", Cambridge Univ. Press, Cambridge, 1990.

[ShTo] G.C. Shephard and J.A. Todd, *Finite unitary reflection groups*, Canad. J. Math. **6** (1954), 274–304.

[St] R. Steinberg, *Differential equations invariant under finite reflection groups*, Trans. Amer. Math. Soc. **112** (1964), 392–400.

Introduction to abelian and derived categories

Bernhard Keller

UFR de Mathématiques, URA 748 du CNRS, Université Paris VII, 2 Place Jussieu, 75251 Paris, France

This is an account of three 1-hour lectures given at the Instructional Conference on Representation Theory of Algebraic Groups and Related Finite Groups, Isaac Newton Institute, Cambridge, 6–11 January 1997.

In section 1, we define abelian categories following Grothendieck [12]. We then characterize module categories among abelian categories. Finally we sketch a proof of Mitchell's full embedding theorem [25]: each small abelian category embeds fully and exactly into a module category.

We come to our main topic in section 2, where we define the derived category of an abelian category following Verdier [33] and the total right derived functor of an additive functor following Deligne [6].

We treat the basics of triangulated categories including K_0-groups and the example of perfect complexes over a ring in section 3.

Section 4 is devoted to Rickard's Morita theory for derived categories [29]. We give his characterization of derived equivalences, list the most important invariants under derived equivalence, and conclude by stating the simplest version of Broué's conjecture [2].

1 Abelian categories

1.1 Definition and basic properties

A **Z**-*category* is a category $\mathcal{C}$ whose morphism sets $\mathrm{Hom}_{\mathcal{C}}(X, Y)$ are abelian groups such that all composition maps

$$\mathrm{Hom}_{\mathcal{C}}(Y, Z) \times \mathrm{Hom}_{\mathcal{C}}(X, Y) \to \mathrm{Hom}_{\mathcal{C}}(X, Z)$$

are bilinear. For example, if R is a ring (associative, with 1) and $\mathcal{C}$ is the category having exactly one object, whose endomorphism set is R, then $\mathcal{C}$ is a **Z**-category. A general **Z**-category should be thought of as a 'ring with several objects' [25].

An *additive category* is a **Z**-category $\mathcal{A}$ which has a zero object 0 (i.e. we have $\mathrm{Hom}_{\mathcal{A}}(0, X) = 0 = \mathrm{Hom}_{\mathcal{A}}(X, 0)$ for all X) and such that all pairs of objects $X, Y \in \mathcal{C}$, admit a *product* in $\mathcal{C}$, i.e. an object $X \prod Y$ endowed with morphisms $p_X : X \prod Y \to X$ and $p_Y : X \prod Y \to Y$ such that the map

$$\mathrm{Hom}_{\mathcal{C}}(U, X\textstyle\prod Y) \to \mathrm{Hom}_{\mathcal{C}}(U, X) \times \mathrm{Hom}_{\mathcal{C}}(U, Y)\,,\; h \mapsto (p_X\, h, p_Y\, h)$$

is bijective. In other words, the pair of maps (p_X, p_Y) is universal among all pairs of morphisms (f, g) from an object U to X and Y, respectively.

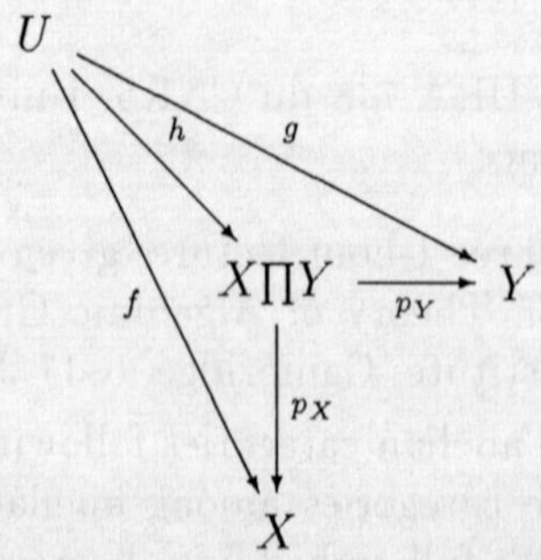

Universal properties of this type are most conveniently expressed in the language of *representable functors*: Recall that a contravariant functor F defined on a category $\mathcal{C}$ with values in the category of sets is *representable* if there is an object $Z \in \mathcal{C}$ and an isomorphism of functors

$$\varphi : \mathrm{Hom}_{\mathcal{C}}(?, Z) \xrightarrow{\sim} F.$$

Note that this determines the object Z uniquely up to canonical isomorphism. For example, the product $X \prod Y$ represents the product functor

$$\mathrm{Hom}_{\mathcal{C}}(?, X) \times \mathrm{Hom}_{\mathcal{C}}(?, Y).$$

Dually, a covariant functor $G : \mathcal{C} \to \mathrm{Sets}$ is *corepresentable* if it is isomorphic to $\mathrm{Hom}_{\mathcal{C}}(Z, ?)$ for some $Z \in \mathcal{C}$.

Accordingly the *coproduct* $X \coprod Y$ is defined to corepresent the functor

$$\mathrm{Hom}_{\mathcal{C}}(X, ?) \times \mathrm{Hom}_{\mathcal{C}}(Y, ?)$$

(if this functor is corepresentable). We leave it to the reader as an exercise to check that in an additive category, the coproduct of any pair of objects exists and is canonically isomorphic to their product. We will henceforth write $X \oplus Y$ for both.

Note that in an additive category, the group law on $\mathrm{Hom}_{\mathcal{C}}(X, Y)$ is determined by the underlying category of $\mathcal{C}$. Indeed, for $f, g \in \mathrm{Hom}_{\mathcal{C}}(X, Y)$, we have the following commutative diagram

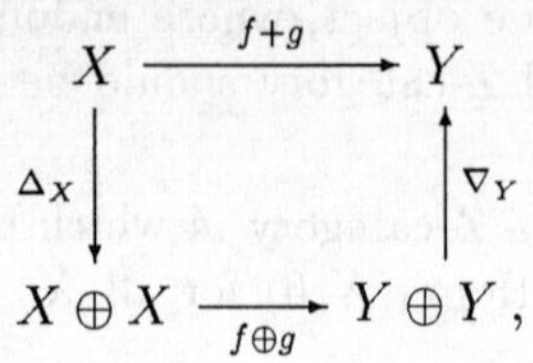

where by definition the composition of the diagonal morphism Δ_X with both of the canonical projections $X \oplus X \to X$ is the identity of X and the codiagonal morphism ∇_Y is defined dually.

If R is a ring, the *category* Mod$R$ of (right) R-modules is an additive category. So are its full *subcategories* Free$R$ *and* mod$R$ whose objects are the free, and the finitely presented R-modules, respectively.

Now let $\mathcal{A}$ be an additive category and $f : A \to B$ a morphism of $\mathcal{A}$. By definition, the *kernel* ker$f$ represents the functor

$$\ker(f_* : \mathrm{Hom}_{\mathcal{A}}(?, A) \to \mathrm{Hom}_{\mathcal{A}}(?, B)).$$

This means that the kernel of f is defined only if this functor is representable, and in this case, the isomorphism from $\mathrm{Hom}_{\mathcal{A}}(?, \ker f)$ to the kernel functor corresponds to a morphism $i : \ker f \to A$ such that $f i = 0$ and i is universal with respect to this property. Dually, the *cokernel* cok$f$ corepresents the functor

$$\ker(f^* : \mathrm{Hom}_{\mathcal{A}}(B, ?) \to \mathrm{Hom}_{\mathcal{A}}(A, ?)).$$

(note that this is the kernel and *not* the cokernel of a morphism between functors). Finally, one defines the *image* $\mathrm{im} f = \ker(B \to \mathrm{cok} f)$ and the *coimage* $\mathrm{coim} f = \mathrm{cok}(\ker f \to A)$. Now suppose that these four objects are well-defined for f. It is then easy to see that there is a unique morphism $\overline{f}$ making the following diagram commutative

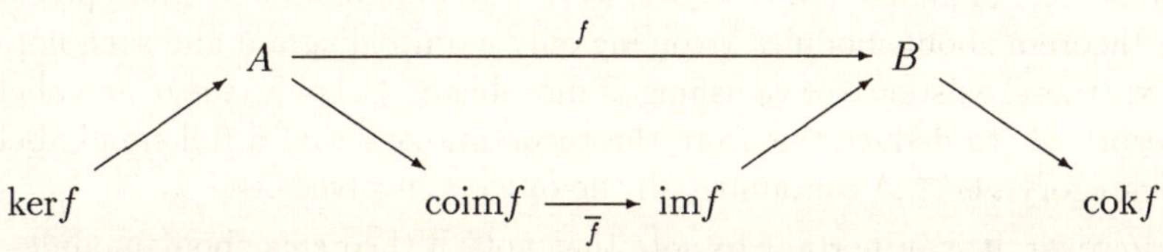

By definition [4] [12], *an abelian category* is an additive category $\mathcal{A}$ such that each morphism of $\mathcal{A}$ admits a kernel and a cokernel and that the canonical morphism $\overline{f}$ is invertible for each morphism f.

This definition implies in particular that in an abelian category a morphism f is invertible iff it is both a monomorphism (i.e. $\ker f = 0$) and an epimorphism (i.e. $\mathrm{cok} f = 0$).

Clearly, if R is a ring, the category Mod$R$ is abelian. If X is a topological space, the category ShX of sheaves of abelian groups on X (cf. [10] [16]) is abelian as well. One of the principal aims of Grothendieck's study [12] of abelian categories was to develop a unified homology theory for these two classes of examples.

It may be helpful to point out two non-examples: If the ring R is not semisimple, the category Proj$R$ of projective modules over R is not abelian since in this case there exist morphisms between projective R-modules which do not admit a cokernel in Proj$R$. But there are also examples of non-abelian categories where each morphism does admit a kernel and a cokernel: This holds for the category of filtered abelian groups

$$A = \bigcup_{n \in \mathbf{N}} A_n.$$

Indeed, if $0 \neq A$, the canonical morphism from A to the filtered group $A(1)$ defined by

$$A(1)_p = A_{p+1}$$

is monomorphic and epimorphic but not invertible.

A functor between abelian categories is *left exact* if it preserves kernels, *right exact* if it preserves cokernels, and *exact* if it is both right and left exact. Recall that a functor $F : \mathcal{A} \to \mathcal{B}$ is said to be *fully faithful* if it induces bijections

$$\mathrm{Hom}_{\mathcal{A}}(A, B) \to \mathrm{Hom}_{\mathcal{B}}(FA, FB)$$

for all objects $A, B \in \mathcal{A}$.

Theorem (Mitchell [24]) *Let $\mathcal{A}_0$ be an abelian category whose objects form a set (i.e. a* small *category). Then there is a ring R and a fully faithful exact functor*

$$F : \mathcal{A}_0 \to \mathrm{Mod}R.$$

This theorem, known as the 'full embedding theorem', allows us to deal with objects of an abelian category 'as if they were modules'. More precisely, any theorem about modules involving only a finite diagram and such notions as exactness, existence or vanishing of morphisms ... holds true in any abelian category $\mathcal{A}$ (to deduce this from the theorem, construct a full small abelian subcategory $\mathcal{A}_0 \subset \mathcal{A}$ containing all the objects involved).

However, it is important to note that not all theorems about module categories carry over to arbitrary abelian categories. For example, the product of an arbitrary set-indexed family of exact sequences of modules is exact; but the analogous statement for sheaves is false, in general. This is not in contradiction with the full embedding theorem, since the functor $\mathcal{A}_0 \to \mathrm{Mod}R$ obtained may not commute with infinite products.

1.2 Characterization of module categories and Morita equivalence

Let R be a ring and $\mathcal{A} = \mathrm{Mod}R$ the category of R-modules. Then it is easy to check that $\mathcal{A}$ has the following properties

- It is *cocomplete*, i.e. for each set-indexed family $(M_i)_{i \in I}$ of objects of $\mathcal{A}$, there exists the coproduct $\bigoplus_{i \in I} M_i$ (which corepresents the functor $\prod_{i \in I} \mathrm{Hom}_{\mathcal{A}}(M_i, ?)$).

- It has a *generator* $P = R$ (the free R-module of rank 1), i.e. for each $M \in \mathcal{A}$, there is an epimorphism $\bigoplus_I P \to M$ for some set I.
- The generator P is *projective*, i.e. the functor $\mathrm{Hom}_{\mathcal{A}}(P, ?) : \mathcal{A} \to \mathrm{Mod}\mathbf{Z}$ is exact.
- The generator P is *compact*, i.e. the functor $\mathrm{Hom}_{\mathcal{A}}(P, ?) : \mathcal{A} \to \mathrm{Mod}\mathbf{Z}$ commutes with arbitrary set-indexed coproducts.

This proves the necessity of the condition of the following

Theorem [7] [8] *Let $\mathcal{A}$ be an arbitrary abelian category and R a ring. Then $\mathcal{A}$ is equivalent to* $\mathrm{Mod}R$ *if and only if $\mathcal{A}$ is cocomplete and has a compact projective generator P with* $\mathrm{Hom}_{\mathcal{A}}(P, P) \cong R$.

To prove the sufficiency, one shows that the functor $F : \mathrm{Hom}_{\mathcal{A}}(P, ?) : \mathcal{A} \to \mathrm{Mod}R$ is an equivalence. In particular, we can take $\mathcal{A}$ to be a module category as well:

Corollary (Morita) *Let R and S be two rings. Then the following conditions are equivalent:*

i) There is an equivalence of categories $F : \mathrm{Mod}R \to \mathrm{Mod}S$.

ii) There is an R-S-bimodule X such that the functor $? \otimes_R X : \mathrm{Mod}R \to \mathrm{Mod}S$ *is an equivalence.*

iii) There is a finitely generated projective S-module P such that P generates $\mathrm{Mod}S$ *and R is isomorphic to* $\mathrm{Hom}_S(P, P)$.

The equivalence between i) and iii) follows from the theorem, once it is shown that a projective S-module P is compact iff it is finitely generated. This is left to the reader as an easy exercise. Clearly ii) implies i). To prove that iii) implies ii), one notes that P has a structure of R-S-bimodule and puts $X = P$. Then it is not hard to verify that ii) holds.

By definition, R is *Morita equivalent* to S if the conditions of the corollary hold. In the best known example, R is the ring of $n \times n$ matrices over S and P is S^n (realized as a set of row vectors on which R acts from the right).

1.3 On the proof of the full embedding theorem

The following sketch of the proof of the full embedding theorem is to give the reader an idea of some more advanced techniques of the theory of abelian categories. We follow Freyd [7].

The proof rests on the following

Theorem (Mitchell) *Let $\mathcal{A}$ be a cocomplete abelian category with a projective generator P. Then each small full abelian subcategory $\mathcal{A}_0 \subset \mathcal{A}$ admits a fully faithful exact functor $F : \mathcal{A}_0 \hookrightarrow$* $\mathrm{Mod}R$ *for some ring R.*

Note that the generator P is *not* supposed to be compact. For the proof, one chooses Q to be a large sum of copies of P; so large indeed that for each object A of $\mathcal{A}_0$ there exists an epimorphism $Q \to A$. Since P is a generator and $\mathcal{A}_0$ is small, this is possible. Now one takes $R = \mathrm{Hom}_{\mathcal{A}}(Q, Q)$ and checks that the restriction F of $\mathrm{Hom}_{\mathcal{A}}(Q, ?)$ to $\mathcal{A}_0$ is fully faithful (cf. [7, Theorem 4.44]).

This proof is still of the same level of difficulty as the proofs of the preceding section. Now, however, we will need some deeper results: As a first trial at 'embedding' $\mathcal{A}_0$, consider the Yoneda embedding

$$Y : \mathcal{A}_0^{op} \to \mathrm{Fun}(\mathcal{A}_0, \mathrm{Mod}\mathbf{Z})\ ,\ A \mapsto \mathrm{Hom}_{\mathcal{A}_0}(A, ?).$$

Here, $\mathrm{Fun}(\mathcal{A}_0, \mathrm{Mod}\mathbf{Z})$ denotes the category of additive functors from $\mathcal{A}_0$ to $\mathrm{Mod}\mathbf{Z}$ (note that this is indeed a category since $\mathcal{A}_0$ is small). Recall that $\mathcal{A}_0$ should be thought of as a 'ring with several objects' and accordingly, $\mathrm{Fun}(\mathcal{A}_0, \mathrm{Mod}\mathbf{Z})$ is viewed as the category of modules over this 'multi-ring'. From this viewpoint, we have already got quite close to our aim of embedding $\mathcal{A}_0$ in a module category. However, the Yoneda functor is *not exact* (only left exact). To remedy this, we observe that the $\mathrm{Hom}_{\mathcal{A}_0}(A, ?)$ are not arbitrary functors : they are *left exact*. We therefore restrict the domain of the Yoneda functor to the category $\mathrm{Lex} = \mathrm{Lex}(\mathcal{A}_0, \mathrm{Mod}\mathbf{Z})$ of left exact functors $\mathcal{A}_0 \to \mathrm{Mod}\mathbf{Z}$

$$Y : \mathcal{A}_0^{op} \to \mathrm{Lex}.$$

The crucial point of the proof is to show that the category of left exact functors is abelian [8]. It is then not hard to see that it is also cocomplete, has a generator (to wit, the direct sum of the functors $\mathrm{Hom}_{\mathcal{A}}(A, ?)$, $A \in \mathcal{A}_0$), and has exact filtered direct limits. In other words, it is a *Grothendieck category* (Grothendieck invented, but did not name, Grothendieck categories in [12]; cf. [28] for a comprehensive account of the subject). Now as a Grothendieck category, the category Lex is also complete and has an injective cogenerator. So we have embedded $\mathcal{A}_0^{op}$ in a complete abelian category with an injective cogenerator. Looking at this through a mirror we see that we have embedded $\mathcal{A}_0$ in Lex^{op}, a cocomplete abelian category with a projective generator. Now we obtain the required embedding $\mathcal{A}_0 \to \mathrm{Mod}R$ from the theorem above.

2 Derived categories and derived functors

Derived categories are a 'formalism for hyperhomology' [34]. Used at first only by the circle around Grothendieck they have now become wide-spread in a number of subjects beyond algebraic geometry, and have found their way into graduate text books [35], [17], [22], [16]. We refer to L. Illusie's account [15] for a brief history of the origins of derived categories.

In order to illustrate the relation between the language of classical homological algebra and that of derived categories, let us consider the example of the Lyndon-Hochschild-Serre spectral sequence: Recall that if G is a group, H a normal subgroup, and A a G-module, then this sequence reads as follows

$$E_2^{pq} = H^p(G/H, H^q(H,A)) \Rightarrow H^{p+q}(G,A). \tag{2.1}$$

The corresponding statement in the language of derived categories is

$$\mathbf{R}\mathrm{Fix}_{G/H} \circ \mathbf{R}\mathrm{Fix}_H = \mathbf{R}\mathrm{Fix}_G \,, \tag{2.2}$$

where the equality denotes a canonical isomorphism between functors defined on the derived category $\mathcal{D}^+\mathrm{Mod}\mathbf{Z}G$ with values in $\mathcal{D}^+\mathrm{Mod}\mathbf{Z}$ and $\mathbf{R}\mathrm{Fix}_G$ the total right derived functor of the fixed point functor $\mathrm{Fix}_G : \mathrm{Mod}\mathbf{Z}G \to \mathrm{Mod}\mathbf{Z}$ defined by

$$\mathrm{Fix}_G M = \{m \in M \mid gm = m \,,\ \forall g \in G\}.$$

Of course, the composition formula (2.2) is based on the observation that $\mathrm{Fix}_{G/H} \circ \mathrm{Fix}_H = \mathrm{Fix}_G$. It is stronger than (2.1) in the sense that (2.1) can be derived from (2.2) by standard techniques [34]. The precise meaning of (2.2) will become clear below. To link the two formulas, we have to evaluate $\mathbf{R}\mathrm{Fix}_G$ at the module A. This is done by applying the functor Fix to an injective resolution

$$I^\cdot : 0 \to I^0 \to I^1 \to \dots$$

of A. By definition, $\mathbf{R}\mathrm{Fix}_G A$ is the complex thus obtained. The link between (2.1) and (2.2) is then the formula

$$H^n\mathbf{R}\mathrm{Fix}_G A = \mathbf{R}^n\mathrm{Fix}_G A = H^n(G,A) \,,$$

where $\mathbf{R}^n\mathrm{Fix}_G$ is the n-th right derived functor of Fix_G in the sense of Cartan-Eilenberg [5].

2.1 Definition of derived categories

Let $\mathcal{A}$ be an abelian category (for example, the category $\mathrm{Mod}R$ of modules over a ring R). We denote by $\mathcal{CA}$ *the category of differential complexes*

$$A^\bullet = (\dots \to A^n \xrightarrow{d_A^n} A^{n+1} \to \dots) \,,\ A^n \in \mathcal{A} \,,\ n \in \mathbf{Z} \,,\ d^2 = 0.$$

Recall that a morphism of complexes $f : A^\bullet \to B^\bullet$ is *null-homotopic* if $f^n = d_B\, h^n + h^{n+1} d_A$ for all $n \in \mathbf{Z}$ for some family of morphisms $h^n : A^n \to B^{n-1}$. Clearly, any composition gfe is null-homotopic if f is null-homotopic. The *homotopy category* $\mathcal{HA}$ has the same objects as $\mathcal{CA}$. Its morphisms from $A^\bullet$ to $B^\bullet$ are the classes of morphisms of complexes $f : A^\bullet \to B^\bullet$ modulo the null-homotopic morphisms.

Note that the homology functor $H^n : \mathcal{CA} \to \mathcal{A}$ induces a well-defined functor $\mathcal{HA} \to \mathcal{A}$. We define a *quasi-isomorphism* to be a morphism s :

$A^\bullet \to A'^\bullet$ of $\mathcal{HA}$ such that the induced morphisms $H^n s : H^n A^\bullet \to H^n A'^\bullet$ are invertible for all $n \in \mathbf{Z}$. We denote by Σ the class of all quasi-isomorphisms. Our aim is to define the derived category $\mathcal{DA}$ as the 'localization' of $\mathcal{HA}$ at the class Σ. Now by construction, $\mathcal{HA}$ is a $\mathbf{Z}$-category (even an additive category), and should be viewed as a 'ring with several objects'. The following lemma states that the analogues of the Ore conditions in the localization theory of rings hold for the class Σ (the assumption that the elements to be made invertible be non-zero divisors is weakened into condition c).

Lemma 1

a) *Identities are quasi-isomorphisms and compositions of quasi-isomorphisms are quasi-isomorphisms.*

b) *Each diagram*

$$A'^\bullet \stackrel{s}{\leftarrow} A^\bullet \stackrel{f}{\rightarrow} B^\bullet \quad (\textit{resp. } A'^\bullet \stackrel{f'}{\rightarrow} B'^\bullet \stackrel{s'}{\leftarrow} B^\bullet)$$

of $\mathcal{HA}$, where s (resp. s') is a quasi-isomorphism, may be embedded into a square

$$\begin{array}{ccc} A^\bullet & \xrightarrow{f} & B^\bullet \\ {\scriptstyle s}\downarrow & & \downarrow{\scriptstyle s'} \\ A'^\bullet & \xrightarrow[f']{} & B'^\bullet \end{array}$$

which commutes in $\mathcal{HA}$.

c) *Let f be a morphism of $\mathcal{HA}$. Then there is a quasi-isomorphism s such that $sf = 0$ in $\mathcal{HA}$ if and only if there is a quasi-isomorphism t such that $ft = 0$ in $\mathcal{HA}$.*

The lemma is proved for example in [17, 1.6.7]. Clearly condition a) would also be true for the pre-image of Σ in the category of complexes. However, for b) and c) to hold, it is essential to pass to the homotopy category. Historically [15], this observation was the main reason for inserting the homotopy category between the category of complexes and the derived category (the latter can also be defined directly as an 'abstract localization' [9] of the category of complexes at the pre-image of Σ).

Now we define [33] the *derived category* $\mathcal{DA}$ to be the localization of the homotopy category at the class of quasi-isomorphisms. This means that the derived category has the same objects as the homotopy category and that morphisms in the derived category from $A^\bullet$ to $B^\bullet$ are given by 'left fractions' "$s^{-1} \circ f$", i.e. equivalence classes of diagrams

$$\begin{array}{ccccc} & & B'^\bullet & & \\ & {\scriptstyle f}\nearrow & & \nwarrow{\scriptstyle s} & \\ A^\bullet & & & & B^\bullet \end{array}$$

where s is a quasi-isomorphism and a pair (f, s) is equivalent to (f', s') iff there is a commutative diagram of $\mathcal{HA}$

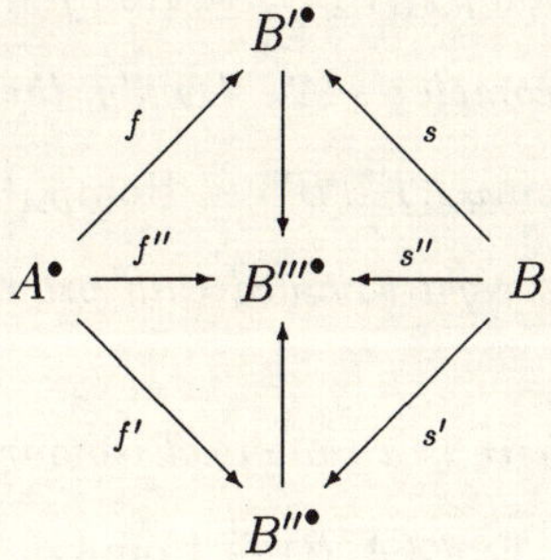

where s'' is a quasi-isomorphism. Composition is defined by

$$“t^{-1}g” \circ “s^{-1}f” = “(s't)^{-1} \circ g'f” ,$$

where $s' \in \Sigma$ and g' are constructed using condition b) as in the following commutative diagram of $\mathcal{HA}$

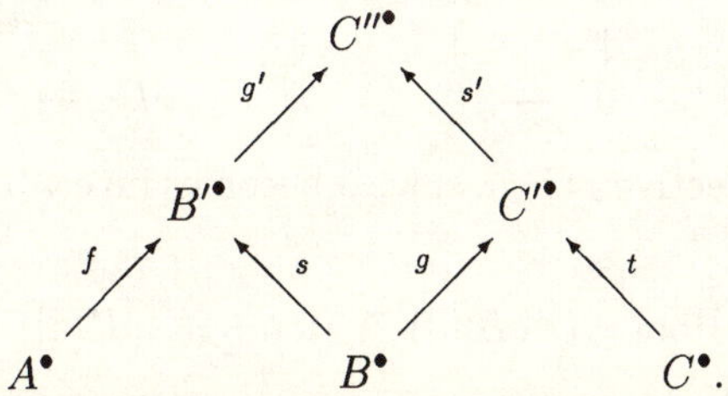

One can then check that composition is associative and admits the obvious morphisms as identities.

Using 'right fractions' instead of left fractions we would have obtained an isomorphic category (use lemma 1 b). We have a canonical functor $\mathcal{HA} \to \mathcal{DA}$ sending a morphism $f : A^\bullet \to B^\bullet$ to the fraction “$\mathbf{1}_B^{-1} f$”. This functor makes all quasi-isomorphisms invertible and is universal among functors with this property. The following lemma yields a more concrete description of some morphisms of the derived category. In part c) we use the following notation: An object $A \in \mathcal{A}$ is identified with the complex $\ldots \to 0 \to A \to 0 \to \ldots$ having A in degree 0. If $K^\bullet$ is an arbitrary complex, we denote by $K^\bullet[n]$ the complex with components $K^\bullet[n]^p = K^{n+p}$ and differential $d_{K[n]} = (-1)^n d_K$.

Lemma 2

a) *The category $\mathcal{DA}$ is additive and the canonical functors $\mathcal{CA} \to \mathcal{HA} \to \mathcal{DA}$ are additive.*

b) *If the complex $I^\bullet$ is left bounded (i.e. $I^n = 0$ for all $n \ll 0$) and has injective components, then the canonical morphism*

$$\mathrm{Hom}_{\mathcal{HA}}(A^\bullet, I^\bullet) \to \mathrm{Hom}_{\mathcal{DA}}(A^\bullet, I^\bullet)$$

is invertible for all complexes $A^\bullet$. Dually, the canonical morphism

$$\mathrm{Hom}_{\mathcal{HA}}(P^\bullet, B^\bullet) \to \mathrm{Hom}_{\mathcal{DA}}(P^\bullet, B^\bullet)$$

is invertible if $P^\bullet$ is right bounded with projective components and $B^\bullet$ is any complex.

c) *For all $A, B \in \mathcal{A}$, there is a canonical isomorphism*

$$\partial : \mathrm{Ext}^n_{\mathcal{A}}(A, B) \xrightarrow{\sim} \mathrm{Hom}_{\mathcal{DA}}(A, B[n]).$$

The calculus of fractions yields part a) of the lemma (cf. [9]). Part b) follows from [14, I, Lemma 4.5]. Part c) is in [14, I, §6].

Let us prove c) in the case where $\mathcal{A}$ has enough injectives (i.e. each object admits a monomorphism into an injective). In this case, the object B admits an injective resolution, i.e. a quasi-isomorphism $s : B \to I^\bullet$ of the form

$$\begin{array}{ccccccccccc} \dots & \to & 0 & \to & B & \to & 0 & \to & 0 & \to & \dots \\ & & \downarrow & & \downarrow & & \downarrow & & \downarrow & & \\ \dots & \to & 0 & \to & I^0 & \to & I^1 & \to & I^2 & \to & \dots \end{array}$$

where the I^p are injective. Then, since s becomes invertible in $\mathcal{DA}$, it induces an isomorphism

$$\mathrm{Hom}_{\mathcal{DA}}(A, B[n]) \xrightarrow{\sim} \mathrm{Hom}_{\mathcal{DA}}(A, I^\bullet[n]).$$

By part b) of the lemma, we have the isomorphism

$$\mathrm{Hom}_{\mathcal{DA}}(A, I^\bullet[n]) \xleftarrow{\sim} \mathrm{Hom}_{\mathcal{HA}}(A, I^\bullet[n]).$$

Finally, the last group is exactly the n-th homology of the complex

$$\mathrm{Hom}_{\mathcal{A}}(A, I^\bullet)\,,$$

which identifies with $\mathrm{Ext}^n_{\mathcal{A}}(A, B)$ by (the most common) definition.

In two very special cases, we can directly describe the derived category in terms of the module category (cf. [17, Exercise I.18]): First suppose that $\mathcal{A} = \mathrm{Mod}k$, where k is a field (or more generally, suppose that $\mathcal{A}$ is semisimple, i.e. $\mathrm{Ext}^1_{\mathcal{A}}(A, B) = 0$ for all $A, B \in \mathcal{A}$). Then the functor $A^\bullet \mapsto H^*A^\bullet$ establishes an equivalence between $\mathcal{DA}$ and the category of $\mathbf{Z}$-graded k-vector spaces. In the second case, suppose that $\mathcal{A}$ is hereditary (i.e. $\mathrm{Ext}^2_{\mathcal{A}}(A, B) = 0$ for all $A, B \in \mathcal{A}$). Then each object $A^\bullet$ of $\mathcal{DA}$ is quasi-isomorphic to the sum of the $(H^n A^\bullet)[-n]$, $n \in \mathbf{Z}$. Morphisms from $A^\bullet$ to $B^\bullet$ are then in bijection with the families (f_n, ε_n), $n \in \mathbf{Z}$, of morphisms $f_n : H^n A^\bullet \to H^n B^\bullet$ and extensions $\varepsilon_n \in \mathrm{Ext}^1_{\mathcal{A}}(H^n A^\bullet, H^{n-1} B^\bullet)$.

2.2 Definition of derived functors

The difficulty in finding a general definition of derived functors is to establish a framework which allows one to derive in full generality as many as possible of the pleasant properties found in the examples. This seems to be best achieved by Deligne's definition [6], which we will give in this section (compare with Grothendieck-Verdier's definition in [33]).

Let $\mathcal{A}$ and $\mathcal{B}$ be abelian categories and $F : \mathcal{A} \to \mathcal{B}$ an additive functor (for example, the fixed point functor $\mathrm{Fix}_G : \mathrm{Mod}\mathbf{Z}G \to \mathrm{Mod}\mathbf{Z}$ from the introduction of this section). Then F clearly induces a functor $\mathcal{CA} \to \mathcal{CB}$ (obtained by applying F componentwise) and a functor $\mathcal{HA} \to \mathcal{HB}$. By abuse of notation, both will be denoted by F as well. We are looking for a functor $? : \mathcal{DA} \to \mathcal{DB}$ so as to make the following square commutative

$$\begin{array}{ccc} \mathcal{HA} & \xrightarrow{F} & \mathcal{HB} \\ \downarrow & & \downarrow \\ \mathcal{DA} & \xrightarrow{?} & \mathcal{DB} \end{array}$$

However, if F is not exact, it will not transform quasi-isomorphisms into quasi-isomorphisms and the functor in question cannot exist. What we will define then is a functor $\mathbf{R}F$ called the 'total right derived functor', which will be a 'right approximation' to an induced functor. More precisely, for a given $A^\bullet \in \mathcal{DA}$, we will not define $\mathbf{R}F(A^\bullet)$ directly but only the functor

$$(\mathbf{r}F)(?, A^\bullet) : (\mathcal{DB})^{op} \to \mathrm{Mod}\mathbf{Z}$$

which, if representable, will be represented by $\mathbf{R}F(A^\bullet)$. For $X^\bullet \in \mathcal{DB}$, we define $(\mathbf{r}F)(X^\bullet, A^\bullet)$ to be the set of 'left F-fractions', i.e. equivalence classes of diagrams

$$\begin{array}{ccccc} & FA'^\bullet & & A'^\bullet & \\ {}^{f}\nearrow & & & & \nwarrow^{s} \\ X^\bullet & & & & A^\bullet, \end{array}$$

where f is a morphism of $\mathcal{DB}$ and s a quasi-isomorphism of $\mathcal{HA}$. Equivalence is defined in complete analogy with section 2.1. We say that $\mathbf{R}F$ is *defined at* $A^\bullet \in \mathcal{DA}$ if the functor $(\mathbf{r}F)(?, A^\bullet)$ is representable and if this is the case, then the value $\mathbf{R}FA^\bullet$ is defined by the isomorphism

$$\mathrm{Hom}_{\mathcal{DB}}(?, (\mathbf{R}F)(A^\bullet)) \xrightarrow{\sim} (\mathbf{r}F)(?, A^\bullet).$$

The link between this definition and more classical constructions is established by the

Proposition *Suppose that $\mathcal{A}$ has enough injectives and $A^\bullet$ is left bounded. Then $\mathbf{R}F$ is defined at $A^\bullet$ and we have*

$$\mathbf{R}FA^\bullet = FI^\bullet$$

where $A^\bullet \to I^\bullet$ is a quasi-isomorphism with a left bounded complex with injective components.

Under the hypotheses of the proposition, the quasi-isomorphism $A^\bullet \to I^\bullet$ always exists [17, 1.7.7]. Viewed in the homotopy category $\mathcal{HA}$ it is functorial in $A^\bullet$ since it is in fact the universal morphism from $A^\bullet$ to a left bounded complex with injective components. For example, if $A^\bullet$ is concentrated in degree 0, i.e. $A^\bullet = A$ for some $A \in \mathcal{A}$, then $I^\bullet$ may be chosen to be an injective resolution of A and we find that

$$H^n \mathbf{R}FA = (\mathbf{R}^n F)(A),$$

the n-th right derived functor of F in the sense of Cartan-Eilenberg [5].

We suggest to the reader as an exercise to prove the identity

$$\mathbf{R}\mathrm{Fix}_{G/H} \circ \mathbf{R}\mathrm{Fix}_H = \mathbf{R}\mathrm{Fix}_G$$

of the introduction of this subsection, where all derived functors are defined on the full subcategory of left bounded complexes $\mathcal{D}^+\mathrm{Mod}\mathbf{Z}G$ of $\mathcal{D}\mathrm{Mod}\mathbf{Z}G$.

3 Triangulated categories

3.1 Definition and examples

Let $\mathcal{A}$ be an abelian category (for example, the category $\mathrm{Mod}R$ of modules over a ring R). One can show that the derived category $\mathcal{DA}$ is abelian only if all short exact sequences of $\mathcal{A}$ split. This deficiency is partly compensated by the so-called triangulated structure of $\mathcal{DA}$, which we are about to define. In this section, to ease the notation, we will write X instead of $X^\bullet$ when speaking of the 'complex X'. Most of the material of this section first appears in [33].

A *standard triangle* of $\mathcal{DA}$ is a sequence

$$X \xrightarrow{Qi} Y \xrightarrow{Qp} Z \xrightarrow{\partial\varepsilon} X[1],$$

where $Q : \mathcal{CA} \to \mathcal{DA}$ is the canonical functor,

$$\varepsilon : 0 \to X \xrightarrow{i} Y \xrightarrow{p} Z \to 0$$

a short exact sequence of complexes, and $\partial\varepsilon$ a certain morphism of $\mathcal{DA}$, functorial in ε, and which lifts the connecting morphism $H^\bullet Z \to H^{\bullet+1}X$ of the long exact homology sequence associated with ε. More precisely, $\partial\varepsilon$ is the fraction "$s^{-1} \circ j$" where j is the inclusion of the subcomplex Z into the complex $X'[1]$ with components $Z^n \oplus Y^{n+1}$ and differential

$$\begin{bmatrix} d_Z & p \\ 0 & -d_Y \end{bmatrix},$$

and $s : X[1] \to X'[1]$ is the morphism $\begin{bmatrix} 0 \\ i \end{bmatrix}$.

A *triangle* of $\mathcal{DA}$ is a sequence (u', v', w') of $\mathcal{DA}$ *isomorphic* to a standard triangle, i.e. such that we have a commutative diagram

$$\begin{array}{ccccccc} X' & \overset{u'}{\to} & Y' & \overset{v'}{\to} & Z' & \overset{w'}{\to} & X'[1] \\ x \downarrow & & \downarrow & & \downarrow & & \downarrow x[1] \\ X & \to & Y & \to & Z & \to & X[1], \end{array}$$

where the vertical arrows are isomorphisms of $\mathcal{DA}$ and the bottom row is a standard triangle.

Lemma 3

T1 For each object X, the sequence

$$0 \to X \overset{1}{\to} X \to 0[1]$$

is a triangle.

T2 If (u, v, w) is a triangle, then so is $(v, w, -u[1])$.

T3 If (u, v, w) and (u', v', w') are triangles and x, y morphisms such that $yu = u'x$, then there is a morphism z such that $zv = v'y$ and $(x[1])w = w'z$.

$$\begin{array}{ccccccc} X & \overset{u}{\to} & Y & \overset{v}{\to} & Z & \overset{w}{\to} & X[1] \\ x \downarrow & & y \downarrow & & z \downarrow & & \downarrow x[1] \\ X' & \overset{u'}{\to} & Y' & \overset{v'}{\to} & Z' & \overset{w'}{\to} & X'[1]. \end{array}$$

T4 For each pair of morphisms

$$X \overset{u}{\to} Y \overset{v}{\to} Z$$

there is a commutative diagram

$$\begin{array}{ccccccc} X & \overset{u}{\to} & Y & \overset{x}{\to} & Z' & \to & X[1] \\ \| & & v \downarrow & & \downarrow w & & \| \\ X & \to & Z & \overset{y}{\to} & Y' & \overset{s}{\to} & X[1] \\ & & \downarrow & & \downarrow t & & \downarrow u[1] \\ & & X' & \overset{1}{\to} & X' & \overset{r}{\to} & Y[1] \\ & & r \downarrow & & \downarrow & & \\ & & Y[1] & \overset{x[1]}{\to} & Z'[1], & & \end{array}$$

where the first two rows and the two central columns are triangles.

Property T4 can be given a more symmetric form if we represent a morphism $X \to Y[1]$ by the symbol $X \overset{+}{\to} Y$ and write a triangle in the form

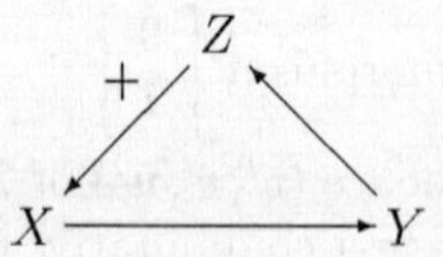

With this notation, the diagram of T4 can be written as an octahedron in which 4 faces represent triangles. The other 4 as well as two of the 3 squares 'containing the center' are commutative.

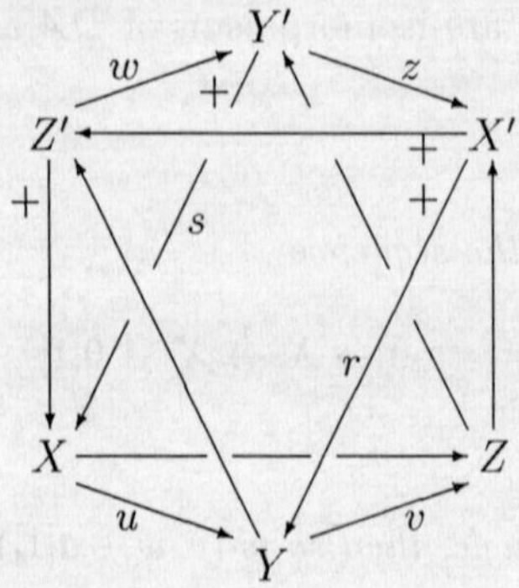

A *triangulated category* is an additive category $\mathcal{T}$ endowed with an auto-equivalence $X \mapsto X[1]$ and a class of sequences (called triangles) of the form

$$X \to Y \to Z \to X[1]$$

which is stable under isomorphisms and satisfies properties T1 through T4.

Note that 'being abelian' is a property of an additive category, whereas 'being triangulated' is the datum of extra structure.

A whole little theory can be deduced from the axioms of triangulated categories. This theory is nevertheless much poorer than that of abelian categories. The main reason for this is the non-uniqueness of the morphism z in axiom T3.

We mention only two consequences of the axioms: a) They are actually self-dual, in the sense that the opposite category $\mathcal{T}^{op}$ also carries a canonical triangulated structure. b) Applying the functor $\mathrm{Hom}_{\mathcal{T}}(U, ?)$ or $\mathrm{Hom}_{\mathcal{T}}(?, V)$ to a triangle yields a long exact sequence of abelian groups. By the 5-lemma, this implies for example that if in axiom T3, two of the three vertical morphisms are invertible, then so is the third.

For later use, we record a number of examples of triangulated categories: If $\mathcal{A}$ is abelian, then not only is the derived category $\mathcal{DA}$ triangulated but also the homotopy category $\mathcal{HA}$. Here the triangles are constructed from componentwise split short exact sequences of complexes.

If $\mathcal{T}$ is a triangulated category, a *full triangulated subcategory of* $\mathcal{T}$ is a full subcategory $\mathcal{S} \subset \mathcal{T}$ such that $\mathcal{S}[1] = \mathcal{S}$ and that whenever we have a triangle (X, Y, Z) of $\mathcal{T}$ such that X and Z belong to $\mathcal{T}$ there is an object Y' of $\mathcal{S}$ isomorphic to Y. For example, the full subcategory $\mathcal{H}^b\mathcal{A}$ of bounded complexes (i.e. $X^p = 0$ for all $|p| \gg 0$) of $\mathcal{HA}$ is a full triangulated subcategory, and so is the full subcategory $\mathcal{D}^b\mathcal{A}$ of bounded complexes of $\mathcal{DA}$. One can show that this subcategory also identifies with the localization of $\mathcal{H}^b\mathcal{A}$ at the class of quasi-isomorphisms between bounded complexes. Note that the categories $\mathcal{HA}$ and $\mathcal{H}^b\mathcal{A}$ are in fact defined for any additive category $\mathcal{A}$.

If $\mathcal{T}$ is a triangulated category and $\mathcal{X}$ a class of objects of $\mathcal{T}$, there is a smallest *strictly* (=closed under isomorphism) full triangulated subcategory $\operatorname{Tria}(\mathcal{X})$ of $\mathcal{T}$ containing $\mathcal{X}$. It is called the *triangulated subcategory generated by* $\mathcal{X}$. For example, the category $\mathcal{D}^b\mathcal{A}$ is generated by $\mathcal{A}$ (identified with the category of complexes concentrated in degree 0).

If R is a ring, a very important triangulated category is the full subcategory $\operatorname{per} R \subset \mathcal{D}\operatorname{Mod} R$ formed by the *perfect* complexes, i.e. the complexes quasi-isomorphic to bounded complexes with components in $\operatorname{proj} R$, the *category* of finitely generated projective R-modules. The subcategory $\operatorname{per} R$ may be intrinsically characterized [29, 6.3] as the subcategory of *compact objects* of $\mathcal{D}\operatorname{Mod} R$, i.e. objects X whose associated functor $\operatorname{Hom}(X, ?)$ commutes with arbitrary set-indexed coproducts. Note that by lemma 2, the canonical functor

$$\mathcal{H}^b\operatorname{proj} R \to \operatorname{per} R$$

is an equivalence so that the category $\operatorname{per} R$ is relatively accessible to explicit computations.

3.2 Grothendieck groups

The *Grothendieck group* $K_0(\mathcal{T})$ of a triangulated category $\mathcal{T}$ is defined [13] as the quotient of the free abelian group on the isomorphism classes $[X]$ of objects of $\mathcal{T}$ divided by the subgroup generated by the relators

$$[X] - [Y] + [Z]$$

where (X, Y, Z) runs through the triangles of $\mathcal{T}$.

For example, if R is a right coherent ring, then the category $\operatorname{mod} R$ of finitely presented R-modules is abelian and the K_0-group of the triangulated category $\mathcal{D}^b\operatorname{mod} R$ is isomorphic to $G_0 R = K_0(\operatorname{mod} R)$ via the Euler characteristic:

$$[M^\bullet] \mapsto \sum_{i \in \mathbf{Z}} (-1)^i [H^i M^\bullet].$$

If R is any ring, the K_0-group of the triangulated category $\operatorname{per} R$ is isomorphic to $K_0 R$ via the morphism

$$[P^\bullet] \mapsto \sum_{i \in \mathbf{Z}} (-1)^i [P^i], \quad P^\bullet \in \mathcal{H}^b\operatorname{proj} R.$$

Note that this shows that any two rings with the 'same' derived category, will have isomorphic K_0-groups. To make this more precise, we need the notion of a triangle equivalence (cf. below)

3.3 Triangle functors

Let $\mathcal{S}, \mathcal{T}$ be triangulated categories. A *triangle functor* $\mathcal{S} \to \mathcal{T}$ is a pair (F, φ) formed by an additive functor $F : \mathcal{S} \to \mathcal{T}$ and a functorial isomorphism

$$\varphi X : F(X[1]) \xrightarrow{\sim} (FX)[1],$$

such that the sequence

$$FX \xrightarrow{Fu} FY \xrightarrow{Fv} FZ \xrightarrow{(\varphi X) Fw} (FX)[1]$$

is a triangle of $\mathcal{T}$ for each triangle (u, v, w) of $\mathcal{S}$.

For example, if $\mathcal{A}$ and $\mathcal{B}$ are abelian categories and $F : \mathcal{A} \to \mathcal{B}$ is an additive functor, one can show [6] that the domain of definition of the right derived functor $\mathbf{R}F$ is a strictly full triangulated subcategory $\mathcal{S}$ of $\mathcal{D}\mathcal{A}$ and that $\mathbf{R}F : \mathcal{S} \to \mathcal{D}\mathcal{B}$ becomes a triangle functor in a canonical way.

A triangle functor (F, φ) is a *triangle equivalence* if the functor F is an equivalence. We leave it to the reader as an exercise to define 'morphisms of triangle functors', and 'quasi-inverse triangle functors', and to show that a triangle functor admits a 'quasi-inverse triangle functor' if and only if it is a triangle equivalence [18].

4 Morita theory for derived categories

The following theorem is the precise analogue of the Morita theorem of section 1.2 in the framework of derived categories.

Let k be a commutative ring. A k-category is a category whose morphism spaces are k-modules such that the composition maps are bilinear (we have already encountered the case $k = \mathbf{Z}$ in section 1.1). A functor between k-categories is *k-linear* if it induces k-linear maps in the morphism spaces.

The following theorem is due to J. Rickard [29] [31]. A direct proof can be found in [21].

Theorem (Rickard) *Let A and B be k-algebras which are flat as modules over k. The following are equivalent*

i) There is a k-linear triangle equivalence $(F, \varphi) : \mathcal{D}\mathrm{Mod}\,A \to \mathcal{D}\mathrm{Mod}\,B$.

ii) There is a complex of A-B-modules $X^\bullet$ such that the total left derived functor

$$\mathbf{L}(? \otimes_A X^\bullet) : \mathcal{D}\mathrm{Mod}\,A \to \mathcal{D}\mathrm{Mod}\,B$$

is an equivalence.

iii) There is a complex T of B-modules such that the following conditions hold

a) T is perfect,

b) T generates $\mathcal{D}\mathrm{Mod}B$ as a triangulated category with infinite direct sums,

c) we have

$$\mathrm{Hom}_{\mathcal{D}B}(T, T[n]) = 0 \textit{ for } n \neq 0 \textit{ and } \mathrm{Hom}_{\mathcal{D}B}(T, T) \cong A \ ;$$

Condition b) in iii) means that $\mathcal{D}\mathrm{Mod}B$ coincides with its smallest strictly full triangulated subcategory stable under forming arbitrary (set-indexed) coproducts.

The implication from ii) to i) is clear. To prove the implication from i) to iii), one puts $T = FA$ (where A is regarded as the free right A-module of rank one concentrated in degree 0). Since F is a triangle equivalence, it is then enough to check that the analogues of a), b), and c) hold for the object A of $\mathcal{D}\mathrm{Mod}A$. Properties a) and c) are clear. Checking property b) is non-trivial [21]. The hard part of the proof is the implication from iii) to ii). Indeed, motivated by the proof of the classical Morita theorem we would like to put $X = T$. The problem is that although A acts on T as an object of the derived category, it does not act on the individual components of T, so that T is not a complex of bimodules as required in ii). We refer to [19] for a direct solution of this problem.

Condition b) of iii) may be replaced by the condition that the direct summands of T generate $\mathrm{per}B$ as a triangulated category, which is easier to check in practice.

If the algebras A and B are even projective as modules over k, then the complex $X^\bullet$ may be chosen to be bounded and with components which are projective from both sides. In this case, the tensor product functor $? \otimes_A X^\bullet$ is exact and induces in the derived category a functor isomorphic to its total left derived functor.

By definition [31], the algebra A is *derived equivalent* to B if the conditions of the theorem hold. In this case, T is called a *tilting complex*, X a *two-sided tilting complex* and $\mathbf{L}(? \otimes_A X)$ a *standard equivalence*.

We know that any equivalence between module categories is given by the tensor product with a bimodule. Strangely enough, in the setting of derived categories, it is an open question whether all k-linear triangle equivalences are (isomorphic to) standard equivalences.

One of the main motivations for considering derived categories is the fact that they contain a large amount of information about classical homological invariants. The following theorem illustrates this point.

Theorem *If A is derived equivalent to B, then*

a) *there is a triangle equivalence* $\mathrm{per}A \xrightarrow{\sim} \mathrm{per}B$ *(and conversely, if there is such an equivalence, then* A *is derived equivalent to* B*);*

b) *if* A *and* B *are right coherent, there is a triangle equivalence* $\mathcal{D}^b\mathrm{mod}A \xrightarrow{\sim} \mathcal{D}^b\mathrm{mod}B$ *(and conversely, if* A *and* B *are right coherent and there is such an equivalence, then* A *is derived equivalent to* B*);*

c) *there is an isomorphism* $K_0A \xrightarrow{\sim} K_0B$ *and, if* A *and* B *are right coherent, an isomorphism* $G_0A \xrightarrow{\sim} G_0B$*;*

d) *the algebras* A *and* B *have isomorphic centers, isomorphic Hochschild homology and cohomology and isomorphic cyclic homology.*

The theorem is proved in [29], [31] and, for the case of cyclic homology, in [20].

A large number of derived equivalent (and Morita non equivalent) algebras is provided by Broué's conjecture [2], [3], which, in its simplest form, is the following statement

Conjecture (Broué) *Let* k *be an algebraically closed field of characteristic* p *and let* G *be a finite group with abelian* p*-Sylow subgroups. Then* $B_{pr}(G)$ *(the principal block of of* kG*) is derived equivalent to* $B_{pr}(N_G(P))$*, where* P *is a* p*-Sylow subgroup.*

We refer to [30] for a proof of the conjecture for blocks of group algebras with cyclic p-Sylow subgroups.

5 Notes on the references

Chapter I of Kashiwara-Schapira's monograph [17] is a concise and very well-written introduction to derived categories (readers may want to consult [14, Chapter I] or [11] to fill in some details). A modern text on homological algebra including derived categories is Weibel's book [35]. Gelfand-Manin [22] give a comprehensive overview of the same subject.

J. Rickard's paper [29] is the original reference for Morita theory for derived categories. The link with derived equivalences is established in [31]. Reference [21] contains direct proofs of the results of [29] and [31].

The articles [32], [26], and [1] by N. Spaltenstein, A. Neeman and M. Boekstedt contain important advances in the treatment of unbounded complexes. These have lead to an improved understanding [27], [23] of the original applications of derived categories in Grothendieck's duality theory [14].

References

[1] M. Bökstedt, A. Neeman, *Homotopy limits in triangulated categories*, Compositio Math. 86 (1993), 209-234.

[2] M. Broué, *Blocs, isométries parfaites, catégories dérivées*, C. R. Acad. Sci. Paris **307** (1988), 13–18.

[3] M. Broué, *Rickard equivalences and block theory*, in: C. M. Campbell (ed.), Groups 93 Galway/St. Andrews, Proceedings, Volume 1, LMS Lecture notes **211**, Cambridge University Press, 1995, 58-79.

[4] D. A. Buchsbaum, *Exact categories and duality*, Trans. AMS **80** (1955), 1–34.

[5] H. Cartan, S. Eilenberg, *Homological Algebra*, Princeton University Press, 1956.

[6] P. Deligne, *Cohomologie à supports propres*, Exposé XVII, SGA 4, Springer LNM **305** (1973), 252-480.

[7] P. Freyd, *Abelian categories*, Harper, 1966.

[8] P. Gabriel, *Des catégories abéliennes*, Bull. Soc. Math. France, **90** (1962), 323-448.

[9] P. Gabriel, M. Zisman, *Calculus of Fractions and Homotopy Theory*, Ergebnisse 35, Springer-Verlag, 1967.

[10] R. Godement, *Théorie des faisceaux*, Hermann, Paris, 1958.

[11] P.-P. Grivel, *Catégories dérivées et foncteurs dérivés*, in A. Borel et al., *Algebraic D-modules*, Perspectives in Mathematics, vol. 2, Academic Press, Orlando, Florida, 1987, 1–108.

[12] A. Grothendieck, *Sur quelques points d'algèbre homologique*, Tôhoku Math. J. **9** (1957), 119–221.

[13] A. Grothendieck, *Groupes de classes des catégories abéliennes et triangulées, Complexes parfaits*, SGA 5, Exposé VIII, Springer LNM **589** (1977), 351-371.

[14] R. Hartshorne, *Residues and duality*, Springer LNM **20**, 1966.

[15] L. Illusie, *Catégories dérivées et dualité: travaux de J.-L. Verdier*, Ens. Math. (2) **36** (1990), 369-391.

[16] B. Iversen, *Cohomology of Sheaves*, Springer, 1986.

[17] M. Kashiwara, P. Schapira, *Sheaves on Manifolds*, Grundlehren 292, Springer-Verlag, Berlin Heidelberg, 1990.

[18] B. Keller, D. Vossieck, *Sous les catégories dérivées*, C. R. Acad. Sci. Paris **305** (1987), 225-228.

[19] B. Keller, *A remark on Tilting theory and DG algebras*, Manus. Math. **79** (1993), 247-252.

[20] B. Keller, *Invariance and localization for cyclic homology of DG algebras*, preprint[1], May 1995, 40 pages, to appear in Journal of Pure and Applied Algebra.

[21] B. Keller, *On the construction of triangle equivalences*, Contribution[1] to: S. König, A. Zimmermann (eds.), 'Derived equivalences of group rings', to appear in the Springer Lecture notes series.

[22] A. I. Kostrikin, I. R. Shafarevich (Eds.) *Algebra V: Homological algebra*, with contributions by S. I. Gelfand and Yu. I. Manin, Encyclopaedia of Mathematical Sciences **38**, Springer, 1994.

[23] J. Lipman, *Notes on derived categories and derived functors*, Preprint (available at http://www.math.purdue.edu/~lipman).

[24] B. Mitchell, *The full imbedding theorem*, Am. J. Math. **86** (1964), 619-637.

[25] B. Mitchell, *Rings with several objects*, Adv. in Math. **8** (1972), 1–161.

[26] A. Neeman, *The Connection between the K-theory localization theorem of Thomason, Trobaugh and Yao and the smashing subcategories of Bousfield and Ravenel*, Ann. Sci. Ecole Norm. Sup. **25** (1992), 547–566.

[27] A. Neeman, *The Grothendieck duality theorem via Bousfield's techniques and Brown representability*, J. Amer. Math. Soc. **9** (1996), 205–236.

[28] N. Popescu, *Abelian categories with applications to rings and modules*, L.M.S. Monographs 3, Academic Press, London, New York.

[29] J. Rickard, *Morita theory for derived categories*, J. London Math. Soc. (2) **39** (1989), 436-456.

[30] J. Rickard, *Derived categories and stable equivalence*, J. Pure and Applied Algebra **61** (1989), 303–317.

[31] J. Rickard, *Derived equivalences as derived functors*, J. London Math. Soc. (2) **43** (1991), 37-48.

[32] N. Spaltenstein, *Resolutions of unbounded complexes*, Compositio Mathematica **65** (1988), 121-154.

[33] J.-L. Verdier, *Catégories dérivées, état 0*, SGA 4 1/2, Springer LNM **569**, 1977, 262-311.

[1]available at www.mathp7.jussieu.fr/ura748/personalpages/keller/recent.html

[34] J.-L. Verdier, Catégories dérivées, Thèse, to appear in Astérisque.

[35] C. A. Weibel, *Introduction to homological algebra*, Cambridge Studies in Advanced Mathematics **38**, Cambridge University Press, 1994.

Finite Groups of Lie Type

Meinolf Geck

UFR de Mathématiques et UMR 9994 du CNRS, Université Paris 7, 2 Place Jussieu, F-75251 Paris

1 Introduction

The starting point of our subject is Chevalley (1955). In this paper, it is shown that one can associate with any field k and any finite dimensional simple Lie algebra over $\mathbb{C}$ a group $G(k)$ which turns out to be simple except possibly if k is finite of very small order. Very roughly the construction goes as follows. Let $\mathfrak{g}$ be a simple Lie algebra over $\mathbb{C}$. Using the Cartan decomposition, one can find a basis $\{e_i, h_j, f_i\}$ of $\mathfrak{g}$ where the h_j span a Cartan subalgebra and the e_i, f_i correspond to positive and negative roots, respectively. Chevalley shows that this can be done so that the structure constants of $\mathfrak{g}$ with respect to this basis are rational integers. Hence the elements e_i, h_j, f_i span a Lie algebra $\mathfrak{g}_{\mathbb{Z}}$ over $\mathbb{Z}$. Tensoring with any field k we thus obtain a Lie algebra $\mathfrak{g}_k$ over k. With any base element e_i or f_i and any element $t \in k$, Chevalley associates an automorphism of $\mathfrak{g}_k$ by imitating the familiar exponential construction over $\mathbb{C}$. These automorphisms generate a subgroup $G(k) \leq \mathrm{Aut}(\mathfrak{g})$.

Thus, it was possible to find a uniform construction for most of the finite simple groups known at the time, and new series of simple groups were discovered this way. Variations of Chevalley's construction yield the twisted groups and the Suzuki and Ree groups. By the classification of finite simple groups in 1981 it is now known that every finite simple group is either cyclic of prime order, or an alternating group of degree 5 or bigger, or a simple group of Lie type as above, or one of 26 sporadic groups. This illustrates the importance of groups of Lie type in finite group theory.

Steinberg (1968) emphasizes a different point of view: a group of Lie type as above over the finite field $\mathbb{F}_q$ can be seen as the fixed point set of a connected reductive algebraic group G over an algebraic closure of $\mathbb{F}_q$ under a Frobenius map F. This makes it possible to use results from the structure theory of connected reductive groups for the study of the finite groups and their representations.

A major breakthrough in understanding the irreducible characters (over an algebraically closed field of characteristic 0) of a finite group of Lie type $G(\mathbb{F}_q)$ was achieved by Deligne and Lusztig (1976) who used the theory of ℓ-adic cohomology to construct families of virtual representations associated with the various F-stable maximal tori in G. The subsequent work of Lusztig

(1984a) led to a complete classification of the irreducible characters of $G(\mathbb{F}_q)$ in terms of parameters in the Langlands dual of G and combinatorial objects attached to the Weyl group of G.

In Sections 1–5 we will describe Frobenius maps on connected reductive algebraic groups and applications of Lang's Theorem. In Section 6 we explain the concept of families of characters of finite Weyl groups. This is deeply related with the Kazhdan-Lusztig theory of cells of finite Weyl groups and the corresponding Iwahori–Hecke algebras. Its relevance to the character theory of finite groups of Lie type will only become apparent in Section 10, after we have explained the Deligne–Lusztig construction of twisted induction in Section 7, the concept of the dual group in Section 8, and Lusztig's Jordan decomposition of characters in Section 9. Finally, in Section 11, we give a brief outline of how Lusztig's theory of character sheaves leads to an explicit algorithm for computing the values of all irreducible characters of $G(\mathbb{F}_q)$.

Several important subjects relevant to the representation theory of finite groups of Lie type have not been discussed in this survey: for example, the theory of Harish-Chandra series and Hecke algebras, both for representations in characteristic 0 and for representations over fields of characteristic $\ell > 0$ where ℓ does not divide q (the "non-defining characteristic" case). We have been content with providing suitable references in Remark 7.2.

2 Fields of definition

Let k be any algebraically closed field. Recall from Carter's lectures that a subset $V \subseteq k^n$ is called algebraic if it is the common set of zeroes of a collection of polynomials in $k[X_1, \ldots, X_n]$. If $V \subseteq k^n$ is such an algebraic subset, it may happen that the coefficients of the polynomials defining V lie in some subfield k_0 of k. In this case we will want to say that "V is defined over k_0" or that "V has a k_0-rational structure". However, some care is needed in this definition since it might depend on the choice of polynomials defining V. We obtain an unambiguous definition as follows.

Definition 2.1 An algebraic subset $V \subseteq k^n$ is defined over a subfield $k_0 \subseteq k$ if the ideal $I(V) \subseteq k[X_1, \ldots, X_n]$ consisting of all polynomials vanishing on V is generated by polynomials in $k_0[X_1, \ldots, X_n]$.

There is still one problem with this definition: we will want to talk about an affine variety V without referring to an embedding of V as an algebraic subset in some k^n. An intrinsic definition should just refer to the coordinate ring $k[V] := k[X_1, \ldots, X_n]/I(V)$ of V. Now, following Steinberg (1974), (1.1), we can characterize the pair $(V, k[V])$ by the following formal properties: (1) $k[V]$ is a finitely generated k-algebra of k-valued functions on V; (2) the map $v \mapsto \lambda_v$ (where λ_v denotes the evaluation at v) defines a bijection between V and the set of all k-algebra homomorphisms from $k[V]$ to k. We are now led to:

Definition 2.2 Let (V, A) be an affine variety, where A is the coordinate ring of V. Then V is defined over a subfield $k_0 \subseteq k$ if there exists a finitely generated k_0-subalgebra $A_0 \subseteq A$ such that the map $k \otimes_{k_0} A_0 \to A$ given by multiplication is an isomorphism.

Recall that a morphism between two affine varieties (V, A) and (W, B) is a map $\varphi \colon V \to W$ such that we have a corresponding co-morphism $\varphi^* \colon B \to A$ given by $f \mapsto f \circ \varphi$. If (V, A) and (W, B) are defined over k_0 it is then clear what it should mean for a morphism to be defined over k_0: let $A = k \otimes_{k_0} A_0$ and $B = k \otimes_{k_0} B_0$; then we require that $\varphi = \mathrm{id} \otimes \varphi_0$ for some k_0-algebra homomorphism $\varphi_0 \colon B_0 \to A_0$.

We will simply write V instead of (V, A) if there is no danger of confusion.

Definition 2.3 Let G be a linear algebraic group, i.e., G is an affine variety and a group such that the multiplication $\mu \colon G \times G \to G$ and the inversion $\iota \colon G \to G$ are morphisms of affine varieties. We say that G is defined over a subfield $k_0 \subseteq k$ if G and the morphisms μ, ι are defined over k_0.

For example, the group $G = \mathrm{GL}_n(k)$ is defined over any subfield of k. This follows from the fact that the coordinate algebra of G is the polynomial ring $k[X_{ij} \mid 1 \leq i, j \leq n]$ localized at $\det = \sum_{\pi \in \mathfrak{S}_n} \mathrm{sgn}(\pi) X_{1\pi(1)} \cdots X_{n\pi(n)}$; clearly, the determinant map, matrix multiplication and inversion are defined over any subfield of k. More generally, the classical groups $\mathrm{Sp}_{2n}(k)$ and $O_n(k)$ are defined over any subfield of k. In fact, this is true for all Chevalley groups over k, whose construction was sketched in the introduction.

For further reading on fields of definition for algebraic groups, see Borel (1991).

3 Frobenius maps

From now on, k will be an algebraic closure of the finite field $\mathbb{F}_q$ where q is a power of a prime p. In this case, we have the following characterization of $\mathbb{F}_q$-rational structures:

Proposition 3.1 *Let (V, A) be an affine variety over k. Then V is defined over $\mathbb{F}_q$ if and only if there exists a k-algebra homomorphism $\gamma \colon A \to A$ such that the following conditions hold:*

(1) We have $\gamma(A) \subseteq A^q$ and the map $\gamma \colon A \to A^q$ is a bijection.

(2) For each $a \in A$ there exists some $m \geq 1$ such that $\gamma^m(a) = a^{q^m}$.

We briefly sketch the proof, but refer to Srinivasan (1979), Chap. II, or Digne and Michel (1991), Prop. 3.3, for the details. If V is defined over $\mathbb{F}_q$ we can find an $\mathbb{F}_q$-subalgebra $A_0 \subseteq A$ such that $A = k \otimes_{\mathbb{F}_q} A_0$, as in Definition 2.2.

Then it can be checked that the map $\gamma\colon A \to A$ defined by $\lambda \otimes a_0 \mapsto \lambda \otimes a_0^q$ ($\lambda \in k$, $a_0 \in A_0$) has the required properties. Conversely, if $\gamma\colon A \to A$ satisfying (1) and (2) is given, then we set $A_0 := \{a \in A \mid \gamma(a) = a^q\}$ which is an $\mathbb{F}_q$-subalgebra of A satisfying the requirements of Definition 2.2.

Definition 3.2 Let (V, A) be an affine variety which is defined over $\mathbb{F}_q$. Let $\gamma\colon A \to A$ be the algebra homomorphism of Proposition 3.1. Then the morphism $F\colon V \to V$ such that $F^* = \gamma$ is called the Frobenius map of V corresponding to the given $\mathbb{F}_q$-rational structure. We shall write $V^F := \{v \in V \mid F(v) = v\}$.

It may be worthwhile to "unfold" the above definition of F. Choosing a finite set of generators for A we can represent A in the form $k[X_1, \ldots, X_n]/I$ for some ideal I, and V can be identified with the algebraic subset of V defined by I. Since V is defined over $\mathbb{F}_q$, generators of A can be chosen in such a way that the ideal I is generated by polynomials in $\mathbb{F}_q[X_1, \ldots, X_n]$. We are now in a situation as in Definition 2.1. Then the map γ is just given by raising the coordinate functions to the q-th power and F is the restriction to V of the map $k^n \to k^n$, $(x_1, \ldots, x_n) \mapsto (x_1^q, \ldots, x_n^q)$. This discussion shows that

$$V^F = \{(x_1, \ldots, x_n) \in V \mid x_i^q = x_i \text{ for all } i\} = V \cap \mathbb{F}_q^n;$$

in particular, V^F is a finite set.

The advantage of the above abstract definition is illustrated, for example, by the following result.

Lemma 3.3 *Let (V, A) be an affine variety defined over $\mathbb{F}_q$, with corresponding Frobenius map F. Let $\sigma\colon V \to V$ be an automorphism of finite order which commutes with F. Then the morphism $F_\sigma = F \circ \sigma$ is also a Frobenius map corresponding to some $\mathbb{F}_q$-rational structure on V.*

Indeed, let γ be the algebra homomorphism corresponding to F as in Def. 3.2. We set $\gamma' := \gamma \circ \sigma^*$. Since σ is an automorphism, condition (1) is certainly satisfied. Let $e \geq 1$ be the order of σ. For any $a \in A$ there exists some $m \geq 1$ such that $\gamma^m(a) = a^{q^m}$. But then we also have $(\gamma')^{me}(a) = \gamma^{me}(a) = a^{q^{me}}$. Hence condition (2) is satisfied and we are done.

Example 3.4 Let $G = \mathrm{GL}_n(k)$ with its "standard" Frobenius map $F\colon G \to G$, $(a_{ij}) \mapsto (a_{ij}^q)$. Then $G^F = \mathrm{GL}_n(q)$, the finite general linear group of non-singular matrices over $\mathbb{F}_q$.

Now consider the map $\sigma\colon G \to G$, $g \mapsto (g^{tr})^{-1}$. This is an automorphism of order 2 which commutes with F. By Lemma 3.3, the map $F_\sigma = F \circ \sigma$ also is a Frobenius morphism. Then G^{F_σ} is the full unitary group over $\mathbb{F}_{q^2}$ with respect to the automorphism $\lambda \mapsto \lambda^q$ of $\mathbb{F}_{q^2}$.

We remark that all finite Chevalley groups as well as the twisted groups of type ${}^2A_n$, ${}^2D_n$, ${}^2E_6$, ${}^3D_4$ arise in this way. In order to obtain the Suzuki and

Ree groups one has to consider groups of the form $G^{F'}$ where $F' \colon G \to G$ is a morphism such that some power of F' is a Frobenius map on G.

We can now state a fundamental theorem of Lang.

Theorem 3.5 (Lang) *Assume that G is a linear algebraic group and that $F \colon G \to G$ is a Frobenius map on G. Then, if G is connected, the map $G \to G$, $g \mapsto g^{-1}F(g)$, is surjective.*

An elegant proof, due to Steinberg, goes as follows. First, the preimage of any element of G under the map $g \mapsto g^{-1}F(g)$ is finite, and hence this map is dominant. It is therefore sufficient to prove that it is a finite morphism. This in turn follows from the finite generation of left and right translates in the coordinate ring of G. For details, we refer to Srinivasan (1979), Theorem 2.4.

4 Applications of Lang's Theorem

We assume that G is a *connected* linear algebraic group and $F \colon G \to G$ is a Frobenius map, corresponding to some $\mathbb{F}_q$-rational structure on G. Furthermore, we assume that G acts transitively on a set X and that there is a map $F' \colon X \to X$ such that the following conditions hold:

(1) The map F' is compatible with F and the action of G on X, i.e., we have $F'(g.x) = F(g).F'(x)$ for all $g \in G$ and $x \in X$.

(2) The stabilizer of a point in X is a closed subgroup of G.

(Note that we do not assume that X is an algebraic variety.)

Let $X^{F'} := \{x \in X \mid F'(x) = x\}$. By condition (1), G^F acts on $X^{F'}$.

Lemma 4.1 *In the above set-up, the set $X^{F'}$ is non-empty.*

Indeed, take any $x \in X$. Since G acts transitively on X, we can find some $g \in G$ so that $F'(x) = g^{-1}.x$. By Lang's Theorem we can write g in the form $g = h^{-1}F(h)$ for some $h \in G$. It is straightforward to check that $h.x \in X^{F'}$, and we are done.

Now let us fix a point $x_0 \in X^{F'}$ and consider its stabilizer $\mathrm{Stab}_G(x_0)$. By condition (2) this is a closed subgroup of G. We shall denote by $A(x_0)$ the quotient of $\mathrm{Stab}_G(x_0)$ by its connected component. Since x_0 is fixed by F' we have an induced action of F on $A(x_0)$ which we denote by the same symbol. We say that $a, a' \in A(x_0)$ are F-conjugate if there exists some $c \in A(x_0)$ such that $a' = c^{-1}aF(c)$. This defines an equivalence relation on $A(x_0)$.

Lemma 4.2 *There is a bijective correspondence between G^F-orbits on $X^{F'}$ and F-conjugacy classes of $A(x_0)$.*

This correspondence is defined as follows. Take any $x \in X^{F'}$. Then there exists some $g \in G$ such that $x = gx_0$. We can now compute that $gx_0 = x = F'(x) = F(g).F'(x_0) = F(g).x_0$ and hence $g^{-1}F(g) \in \mathrm{Stab}_G(x_0)$. Let a be the image of $g^{-1}F(g)$ in $A(x_0)$. If we take another element x' in the same G^F-orbit as x and choose a corresponding element $g' \in G$ we obtain an element $a' \in A(x_0)$ which is F-conjugate to a. Thus, we have defined a map from G^F-orbits on $X^{F'}$ to F-conjugacy classes of $A(x_0)$. It then has to be checked that this map is injective and surjective; for details see E 2.7 in the article of Springer and Steinberg in Borel et al. (1970).

Example 4.3 Let C be an F-stable conjugacy class in G, and $F' \colon C \to C$ be the restriction of F to C. Then conditions (1) and (2) are satisfied; we have $\mathrm{Stab}_G(x) = C_G(x)$ for $x \in C$. We conclude that there exists some $x_0 \in C^F$ and that, moreover, the G^F-classes contained in C^F are in bijection to the F-conjugacy classes of $A(x_0)$. In particular, this shows that if $C_G(x_0)$ is connected then C^F is a single G^F-conjugacy class.

This situation occurs, for example, in $G = \mathrm{GL}_n(k)$ where the centralizer of any element is connected. Another example is given in Lemma 9.1.

Example 4.4 Let X be the set of all pairs (T, B) where $T \subseteq G$ is a maximal torus and $B \subseteq G$ is a Borel subgroup such that $T \subseteq B$. The group G acts transitively by conjugation on X, and we let $F' \colon (T, B) \mapsto (F(T), F(B))$. Then conditions (1) and (2) are satisfied; we have $\mathrm{Stab}_G(T, B) = N_G(B) \cap N_G(T) = B \cap N_G(T) = N_B(T) = C_B(T)$ which is even connected. Hence there exists some F-stable pair (T_0, B_0) and any two such pairs are conjugate by an element of G^F.

Example 4.5 Let X be the set of all maximal tori of G. Then G acts transitively by conjugation and we let $F' \colon T \mapsto F(T)$. The conditions (1) and (2) are satisfied; we have $\mathrm{Stab}_G(T) = N_G(T)$ and $\mathrm{Stab}_G(T)^\circ = C_G(T)$. Thus, we have a bijection between G^F-classes of F-stable maximal tori in G and F-conjugacy classes of W, where $W = N_G(T_0)/C_G(T_0)$ for some fixed F-stable maximal torus T_0. If T lies in the G^F-class of maximal tori parametrized by the element $w \in W$ we shall also write $T = T_w$ to simplify notation.

5 Finite reductive groups

Let G be a connected reductive group and $F \colon G \to G$ be a Frobenius map corresponding to some $\mathbb{F}_q$-rational structure on G. Then the finite group G^F is called a “finite group of Lie type” or a “finite reductive group”. (Note that G^F is indeed a group since F is a bijective homomorphism of algebraic groups.) We shall now explain two important structural properties of these groups.

5.1 Complete root data

Recall from Carter's lectures that the algebraic group G is determined by its root datum with respect to a maximal torus. There is a similar statement taking also into account the $\mathbb{F}_q$-rational structure on G. To describe this, we need to prepare some notation.

As in Example 4.4, we can find an F-stable pair (T_0, B_0) consisting of a maximal torus T_0 and a Borel subgroup $B_0 \supseteq T_0$. Let $W := N_G(T_0)/T_0$ be the corresponding Weyl group and $S \subseteq W$ the set of simple reflections determined by B_0. Then F induces an automorphism of finite order on W which stabilizes S.

Let X, Y be the character and cocharacter group of T_0, respectively, $\Phi \subset X$ the roots and $\Phi^\vee \subset Y$ the coroots. Then F acts on X and Y by:

$$F(\chi)(t) = \chi(F(t)) \quad \text{and} \quad F(\gamma)(\lambda) = F(\gamma(\lambda))$$

where $\chi \in X$, $\gamma \in Y$, $t \in T$ and $\lambda \in k^\times$. These actions leave the sets Φ and $\Phi^\vee$ invariant. The action of F on X can be written in the form $F = qF_0$ where $F_0\colon X \to X$ is an automorphism of finite order with $F_0(\Phi) = \Phi$. The pair (G, F) is then determined by the root datum $(X, \Phi, Y, \Phi^\vee)$, the automorphism F_0 and the prime power q. These ideas have been formalized in the concept of a *complete root datum* by Broué and Malle, see the article by the same authors in this volume.

5.2 Bruhat decomposition

The structure of the algebraic group G is to a large extent controlled by the pair $(B_0, N_G(T_0))$ which forms a Tits system in G. For example, we have the Bruhat decomposition $G = \coprod_{w \in W} B_0 \dot{w} B_0$ where $\dot{w} \in N_G(T_0)$ denotes a fixed representative of $w \in W$. Since we have chosen B_0, T_0 to be F-stable, these representatives can be chosen to be F-stable, too. We then also have a Bruhat decomposition in the finite group G^F:

$$G^F = \coprod_{w \in W^F} B_0^F \dot{w} B_0^F.$$

There is in fact a sharp form of this decomposition which allows us to derive a formula for the order of G^F. We write $B_0 = U_0 T_0$ where U_0 is the unipotent radical of B_0. For each $w \in W$ we denote $U_w = U_0 \cap U_0^{w_0 w}$ where $w_0 \in W$ is the unique element of maximal length in W, cf. Rouquier's lectures. (Note that U_w does not depend on the choice of representative $\dot{w}$.) Then we have $B_0 \dot{w} B_0 = U_0 T_0 \dot{w} U_w$ with uniqueness of expressions. Thus, we conclude that

$$|G^F| = \sum_{w \in W^F} |B_0^F \dot{w} B_0^F| = |U_0|^F |T_0^F| \sum_{w \in W^F} |U_w^F|.$$

Now we have $|U_w^F| = q^{l(w)}$ for $w \in W^F$; in particular, $|U_0^F| = |U_{w_0}^F| = q^{l(w_0)} = q^N$ where N denotes the number of positive roots of Φ. Furthermore, the order of T_0^F can be written as the product of $|Z^\circ(G)^F|$ (where $Z^\circ(G)$ is the connected center of G) and terms of the form $q^{|J|} - 1$ where J runs over the F-orbits on S. Hence we obtain the following formula:

$$|G^F| = q^N |Z^\circ(G)^F| \prod_{J \in S/F} (q^{|J|} - 1) \sum_{w \in W^F} q^{l(w)}.$$

For more details we refer to Carter (1985), §2.9.

6 Characters of finite Weyl groups

Before we can go into the character theory of finite reductive groups we need some basic results about the characters of finite Weyl groups. Our aim is to describe the partition of the set of irreducible characters into *families*, following Lusztig (1984a), Chapter 4.

6.1 Generic degrees

Let W be a finite Weyl group and $S \subset W$ a set of simple reflections of W. Let H be the Iwahori-Hecke algebra associated with (W, S), see Rouquier's lectures. For our purposes here, we assume that H is defined over the field of rational functions $\mathbb{Q}(u^{1/2})$ in an indeterminate $u^{1/2}$ so that the standard basis elements T_w $(w \in W)$ satisfy the relations:

$$\begin{aligned} T_w T_{w'} &= T_{ww'} && \text{if } l(ww') = l(w) + l(w'), \\ T_s^2 &= uT_1 + (u-1)T_s && \text{for } s \in S. \end{aligned}$$

With every irreducible character $\varphi \in \operatorname{Irr}(W)$ one can canonically associate an irreducible character $\varphi_u \in \operatorname{Irr}(H)$, see (3.3) in Lusztig (1984a). The values of φ_u at basis elements T_w lie in $\mathbb{Z}[u^{1/2}]$ and φ is recovered from φ_u by setting $u^{1/2} \mapsto 1$. It was shown by Benson and Curtis (1972) that for each $\varphi \in \operatorname{Irr}(W)$ there exists a polynomial $D_\varphi(u) \in \mathbb{Q}[u]$ such that

$$\sum_{w \in W} u^{-l(w)} \varphi_u(T_w) \varphi_u(T_{w^{-1}}) = \frac{\sum_{w \in W} u^{l(w)}}{D_\varphi(u)} \varphi(1).$$

This polynomial is called the *generic degree* of $\varphi \in \operatorname{Irr}(W)$.

6.2 Truncated induction

Following (4.1) in Lusztig (1984a), we associate with every irreducible character $\varphi \in \operatorname{Irr}(W)$ the integer $a_\varphi \geq 0$ such that $D_\varphi(u) = f_\varphi^{-1} u^{a_\varphi} +$ higher powers of u, where $f_\varphi \neq 0$. It was shown by Lusztig that the assignment $\varphi \mapsto a_\varphi$ has the following property: let $S' \subseteq S$ be a subset and $W' \subseteq W$ the corresponding parabolic subgroup. For $\psi \in \operatorname{Irr}(W')$ and $\varphi \in \operatorname{Irr}(W)$ let $n_{\psi,\varphi}$ be the multiplicity of $\varphi \in \operatorname{Irr}(W)$ in the induced character $\operatorname{Ind}_{W'}^{W}(\psi)$. Then we have $a_\varphi \geq a_\psi$ if $n_{\psi,\varphi} \neq 0$, and given ψ there is at least some φ such that $n_{\psi,\varphi} \neq 0$ and $a_\psi = a_\varphi$. A "truncated" induction is then defined by

$$J_{W'}^{W}(\psi) = \sum_{\varphi} n_{\psi,\varphi}\varphi, \qquad \text{sum over all } \varphi \in \operatorname{Irr}(W) \text{ with } a_\varphi = a_\psi.$$

We extend it by linearity to all characters of W. If W'' is another standard parabolic subgroup such that $W' \subseteq W'' \subseteq W$, we have $J_{W'}^{W} = J_{W''}^{W} \circ J_{W'}^{W''}$.

Definition 6.3 (Lusztig) A character χ of W is called constructible if it satisfies:

(i) If $W = \{1\}$ then χ is the trivial character.

(ii) If $W \neq \{1\}$ then there exists a proper parabolic subgroup $W' \subset W$ and a constructible character χ' of W' such that either $\chi = J_{W'}^{W}(\chi')$ or $\chi \otimes \operatorname{sgn} = J_{W'}^{W}(\chi')$.

The families of $\operatorname{Irr}(W)$ can now be defined as the equivalence classes of the relation generated by the condition "$\varphi, \varphi' \in \operatorname{Irr}(W)$ both occur as constituents of some constructible character".

Example 6.4 (a) Let (W, S) be of type A_{n-1} so that W is isomorphic to $\mathfrak{S}_n$. Then every $\varphi \in \operatorname{Irr}(W)$ is of the form $J_{W'}^{W}(\operatorname{sgn})$ for a suitable parabolic subgroup $W' \subseteq W$ and, conversely, the J-operation always maps irreducible characters to irreducible characters. It follows that each family just consists of one irreducible character.

(b) Let (W, S) be of type B_2. Then W is isomorphic to the dihedral group of order 8. We denote the 5 irreducible characters of W by 1_W, ε, r, ε_1 and ε_2, where 1_W is the trivial character, ε is the sign character, r is the reflection character and $\varepsilon_1, \varepsilon_2$ are two further characters of degree 1. The corresponding generic degrees are 1, u^4, $u(u+1)^2/2$, $u(u^2+1)/2$ and $u(u^2+1)/2$, respectively. A direct computation shows that 1_W, ε, $r+\varepsilon_1$ and $r+\varepsilon_2$ are the constructible characters of W. Hence the families are $\{1\}$, $\{\varepsilon\}$, $\{r, \varepsilon_1, \varepsilon_2\}$.

The constructible characters and families have been determined explicitly by Lusztig for all types; the results can be found in Carter (1985), Chapter 13. From these explicit results one can check, for example, that the map $\varphi \mapsto a_\varphi$ is constant on families.

Barbasch and Vogan have shown (see Theorem 5.25 in Lusztig (1984a)) that two irreducible characters of W lie in the same family if and only if they belong to the same two-sided cell, in the sense defined by Kazhdan and Lusztig. For further reading on this subject we refer to Lusztig (1987).

7 Twisted induction

We let again G be a connected reductive group and $F\colon G \to G$ a Frobenius map corresponding to some $\mathbb{F}_q$-rational structure on G. We now turn to the problem of constructing and classifying irreducible characters of G^F.

7.1 The construction of Deligne and Lusztig

Let $L \subseteq G$ be a closed F-stable subgroup. We say that L is *regular* if L is a Levi complement in some (not necessarily F-stable) parabolic subgroup $P \subseteq G$. In this situation, Deligne and Lusztig have found a generalization of the classical induction due to Frobenius, by using the ℓ-adic cohomology with compact support of certain subvarieties of G on which the finite groups L^F and G^F act.

This construction goes as follows. We write $P = U_P L$ where U_P is the unipotent radical of P and set

$$X_{L,P} := \{x \in G \mid x^{-1}F(x) \in U_P\}.$$

This is an algebraic subset of G on which G^F and L^F act by left and right multiplication, respectively. Indeed, if $g \in G^F$ and $x \in X_{L,P}$ then $(gx)^{-1}F(gx) = x^{-1}F(x) \in U_P$, and hence $gx \in X_{L,P}$; on the other hand, if $l \in L^F$ and $x \in X_{L,P}$ then $(xl)^{-1}F(xl) = l^{-1}x^{-1}F(x)l \in U_P$ since $x^{-1}F(x) \in U_P$ and L normalizes U_P. So $xl \in X_{L,P}$.

Now consider the ℓ-adic cohomology with compact support of $X_{L,P}$, where ℓ is a prime not dividing q; see, for example, Milne (1980). For each $i \geq 0$ we obtain a finite-dimensional $\bar{\mathbb{Q}}_\ell$-vector space $H_c^i(X_{L,P}, \bar{\mathbb{Q}}_\ell)$. The actions of L^F and G^F on $X_{L,P}$ induce a (G^F, L^F)-bimodule structure on these vector spaces. Thus, given a finite-dimensional L^F-module V we can define a virtual G^F-module

$$R_{L,P}^G(V) := \sum_{i \geq 0} (-1)^i H_c^i(X_{L,P}, \bar{\mathbb{Q}}_\ell) \otimes_{\bar{\mathbb{Q}}_\ell L^F} V.$$

Taking characters and extending by linearity yields a map

$$R_{L,P}^G\colon \mathbb{Z}\operatorname{Irr}(L^F) \to \mathbb{Z}\operatorname{Irr}(G^F) \qquad \text{“twisted induction”}.$$

For each i, the dual space of $H_c^i(X_{L,P}, \bar{\mathbb{Q}}_\ell)$ is an (L^F, G^F)-bimodule. In an analogous way as above, we then obtain a map

$$^*R_{L,P}^G\colon \mathbb{Z}\operatorname{Irr}(G^F) \to \mathbb{Z}\operatorname{Irr}(L^F) \qquad \text{“twisted restriction”},$$

which is adjoint to $R_{L,P}^G$ with respect to the usual scalar product for class functions on G^F and L^F, respectively. Recall that this is given by $\langle f, f' \rangle = (1/|G^F|)\sum_{g \in G^F} f(g)f'(g^{-1})$ for class functions f, f' on G^F (and similarly for class functions on L^F).

Remark 7.2 In the above set-up, assume that P is also F-stable. Then we may regard an L^F-module V as a module for P^F with U_P^F acting trivially. Denoting this module by $\tilde{V}$, it can be shown that

$$R_{L,P}^G(V) = \mathrm{Ind}_{P^F}^{G^F}(\tilde{V}).$$

Note that $R_{L,P}^G(V)$ is an actual G^F-module. Thus, the operation of twisted induction reduces to the operation of *Harish-Chandra induction* in this case. There is a corresponding theory of Harish-Chandra series and Hecke algebras, of which an up to date account can be found in Digne and Michel (1991) and Carter (1985). For generalizations or extensions of these ideas see the course by Broué and Malle, and Geck and Hiss (1996) and the references there.

There is a number of properties of the operation of twisted induction which is believed to hold in general, like independence of the choice of the parabolic subgroup, a Mackey formula etc. At present, these properties can only be proved using some (very mild) additional assumptions on G, on V, or on p, q. The situation becomes better if we restrict to the case where $L = T$ is an F-stable maximal torus.

Theorem 7.3 (Scalar product formula) *Let T, T' be F-stable maximal tori, contained in Borel subgroups B, B' of G, respectively. Then we have*

$$\langle R_{T,B}^G(\theta), R_{T',B'}^G(\theta')\rangle = \frac{1}{|T^F|}|\{n \in G^F \mid nTn^{-1} = T' \text{ and } {}^n\theta = \theta'\}|$$

for all irreducible characters $\theta \in \mathrm{Irr}(T^F)$ and $\theta' \in \mathrm{Irr}(T'^F)$.

For the proof, see the original article Deligne and Lusztig (1976) or Section 7.3 in Carter (1985).

One consequence of this formula is that $R_{T,B}^G(\theta)$ is in fact independent of the choice of B. Indeed, assume that T is contained in two Borel subgroups B, B'. Then the above formula shows that

$$\langle R_{T,B}^G(\theta), R_{T,B}^G(\theta)\rangle = \langle R_{T,B}^G(\theta), R_{T,B'}^G(\theta)\rangle = \langle R_{T,B'}^G(\theta), R_{T,B'}^G(\theta)\rangle.$$

This implies that $R_{T,B}^G(\theta) - R_{T,B'}^G(\theta)$ has norm 0, and hence is 0. Therefore, we can omit the subscript B and simply write $R_{T,\theta}^G$ instead of $R_{T,B}^G(\theta)$.

Another consequence is that $\pm R_{T,\theta}^G$ is an irreducible character if $\theta \in \mathrm{Irr}(T^F)$ is in *general position*, i.e., if no non-identity element in $N_G(T)^F/T^F$ fixes θ.

Further relations and properties of the generalized characters $R_{T,\theta}^G$ can be found in Sections 7.4 and 7.5 in Carter (1985).

A class function f on G^F is called *uniform* if it is a linear combination of generalized characters $R_{T,\theta}^G$, for various T, θ. In general, the uniform class functions do not span the space of all class functions on G^F. But, at least, we have the following result:

Theorem 7.4 (Uniformity of the regular representation) *The character of the regular representation of G^F is given by*

$$\chi_{reg} = \frac{1}{|W|} \sum_{w \in W} \sum_{\theta \in \mathrm{Irr}(T_w^F)} R_{T_w,\theta}^G(1) R_{T_w,\theta}^G,$$

where T_w denotes an F-stable maximal torus in the F-conjugacy class of $w \in W$, cf. Example 4.5.

For a proof, see the original article Deligne and Lusztig (1976) or Corollary 12.14 in Digne and Michel (1991). This formula shows that each irreducible character of G^F occurs with non-zero multiplicity in some generalized character $R_{T,\theta}^G$. An analysis of these multiplicities has led Lusztig to a classification of the irreducible characters of G^F.

8 The dual group

We keep our standard setting: G is a connected reductive group defined over $\mathbb{F}_q$ with corresponding Frobenius map F. Let T_0, B_0, W, S be as in (5.1).

Now suppose we have another group G^* defined over $\mathbb{F}_q$ with corresponding Frobenius map F^*, and that T_0^*, B_0^*, W^*, S^* have analogous meanings as before. We then say that G and G^* are dual groups if there is an isomorphism $X(T_0) \xrightarrow{\sim} Y(T_0^*)$ which takes the simple roots in $X(T_0)$ to simple coroots in $Y(T_0^*)$ and which is compatible with the actions of F and F^* on $X(T_0)$ and $Y(T_0^*)$. If this is the case, there is a canonical isomorphism $\delta\colon W \to W^*$ which has the property that

$$\delta(F(w)) = F^{*-1}(\delta(w)) \quad \text{for all } w \in W.$$

Thus, dual groups have isomorphic Weyl groups but the underlying root systems are dual in the sense that the roles of long and short roots are interchanged. Moreover, if G is a simple group of adjoint type then G^* will be simple of simply-connected type. Thus, for example, we have $\mathrm{PGL}_n^* = \mathrm{SL}_n$, $\mathrm{SO}_{2n+1}^* = \mathrm{Sp}_{2n}$, $E_8^* = E_8$. It can also be easily seen that $\mathrm{GL}_n^* = \mathrm{GL}_n$.

The following result shows that one might think of a duality as having the effect of replacing the characters of a torus by the elements in a dual torus.

Lemma 8.1 *Suppose G, F and G^*, F^* are dual as above. Then there is a natural bijective correspondence between*

(a) the set of G^F-conjugacy classes of pairs (T, θ) where $T \subseteq G$ is an F-stable maximal torus and $\theta \in \mathrm{Irr}(T^F)$, and

(b) the set of G^{*F^*}-conjugacy classes of pairs (T^*, s) where $T^* \subseteq G^*$ is an F^*-stable maximal torus and $s \in T^{*F^*}$.

The proof uses the following ingredients. Take an F-stable maximal torus $T \subseteq G$. We can write $T = T_w$ for some $w \in W$, as in Example 4.5. Then let $T^* \subseteq G^*$ be an F^*-stable maximal torus such that $T^* = T^*_{\delta(w)^{-1}}$. This correspondence defines a natural bijection between G^F-classes of F-stable maximal tori in G and G^{*F^*}-conjugacy classes of F^*-stable maximal tori in G^*.

In order to take also into account characters of T^F, we interpret these as elements of $X(T)$. Indeed, choosing an embedding $k^\times \hookrightarrow \bar{\mathbb{Q}}_\ell^\times$, we can define a map $X(T) \to \operatorname{Irr}(T^F)$ by restricting a rational character $\chi\colon T \to k^\times$ to T^F and then composing it with that embedding. It turns out that this gives rise to an isomorphism $\operatorname{Irr}(T^F) \cong X(T)/(F-1)X(T)$.

Dually, we can also relate elements in T^{*F^*} with cocharacters in $Y(T^*)$. In fact, we have an isomorphism $T^{*F^*} \cong Y(T^*)/(F^*-1)Y(T^*)$ which now depends on the choice of an isomorphism between $k^\times$ and $\mathbb{Q}_{p'}/\mathbb{Z}$, where $\mathbb{Q}_{p'}$ is the additive group of rational numbers of the form r/s with $r, s \in \mathbb{Z}$ and s not divisible by p. Since $X(T)$ and $Y(T^*)$ are isomorphic, we conclude that

$$\operatorname{Irr}(T^F) \cong T^{*F^*} \quad \text{for dual tori } T \text{ and } T^*.$$

For details see the original article Deligne and Lusztig (1976), or Section 3.2 in Carter (1985), or Proposition 13.13 in Digne and Michel (1991).

If (T,θ) and (T^*,s) are in duality as in Lemma 8.1, we also write $R^G_{T^*,s}$ instead of $R^G_{T,\theta}$. We can now state the following fundamental result:

Theorem 8.2 (Deligne and Lusztig (1976)) *Two generalized characters $R^G_{T_1^*,s_1}$ and $R^G_{T_2^*,s_2}$ have no irreducible constituent in common unless s_1, s_2 are conjugate in G^*.*

We can thus define a partition of $\operatorname{Irr}(G^F)$ into "Lusztig series"

$$\operatorname{Irr}(G^F) = \coprod_{(s)} \mathcal{E}_{(s)}$$

where (s) runs over the F^*-stable conjugacy classes of semisimple elements in G^*. An irreducible character ρ lies in $\mathcal{E}_{(s)}$ if and only if ρ has non-zero scalar product with $R^G_{T^*,s}$ for some F^*-stable maximal torus $T^* \subseteq G^*$. It can even be shown that, in the above theorem, s_1, s_2 must be conjugate in G^{*F^*}; see (7.5.2) in Lusztig (1977), and also Proposition 14.41 in Digne and Michel (1991).

9 The Jordan decomposition of characters

One main idea in Lusztig's classification of irreducible characters of G^F is to compare the series $\mathcal{E}_{(s)}$ with the series $\mathcal{E}_{(1)}$ defined with respect to the group $C_{G^*}(s)$ — if this makes sense. This is guaranteed by the following result:

Lemma 9.1 *Suppose that G, G^* are dual to each other. If G has a connected center then $C_{G^*}(s)$ is a connected reductive group, for every semisimple element $s \in G^*$.*

The fact that $C_{G^*}(s)^\circ$ is reductive follows from the sharp form of the Bruhat decomposition, see Theorem 3.5.3 in Carter (1985). The assumption that G has a connected center implies, by duality, that there is a bijective homomorphism of algebraic groups between the derived subgroup of G^* and the semisimple simply-connected group of the same type. The connectedness of $C_{G^*}(s)$ in this situation is one of the basic results of Steinberg (1968), see also Remarks 13.15 in Digne and Michel (1991) for an alternative proof.

Definition 9.2 An irreducible character of G^F is called *unipotent* if it lies in the series $\mathcal{E}_{(1)}$, i.e., if it occurs with non-zero multiplicity in $R_{T,1}^G$, for some F-stable maximal torus $T \subseteq G$ (and where 1 stands for the trivial character). The set of unipotent characters will be denoted by $\mathcal{U}(G^F)$.

The unipotent characters play a special role in the representation theory of finite groups of Lie type. It is only in groups of untwisted type A_n where we have an elementary definition for them: in this case they are the constituents of the permutation representation of G^F on B^F, where $B \subseteq G$ is an F-stable Borel subgroup.

The unipotent characters of G^F are "insensitive to the center of G", see Proposition 7.10 in Deligne and Lusztig (1976). Even more is true: the unipotent characters only depend on the type of W as a Coxeter group (and the action of F on it), but not on the underlying root system:

Theorem 9.3 (Lusztig (1984a)) *There is a bijection $\mathcal{U}(G^F) \leftrightarrow \mathcal{U}(G^{*F^*})$, $\rho \leftrightarrow \rho^*$, such that*

$$\langle \rho, R_{T,1}^G \rangle = \langle \rho^*, R_{T^*,1}^{G^*} \rangle$$

where $T \subseteq G$ and $T^* \subseteq G^*$ are F-stable maximal tori dual to each other.

This is proved as follows. First, by using arguments similar to those in (8.8) in Lusztig (1984a), one can reduce to the case where G and G^* are simple algebraic groups defined over $\mathbb{F}_q$. The only case where we have to prove something is when G has type B_n or C_n. But now the results in Chap. 9 in Lusztig (1984a) show that the above theorem holds.

We can now state:

Theorem 9.4 (Lusztig (1984a)) *Assume that the center of G is connected. Let $s \in G^{*F^*}$ be semisimple and $H := C_{G^*}(s)$, which is a connected reductive group defined over $\mathbb{F}_q$. Then there exists a sign $\varepsilon_{(s)}$ and a bijection $\mathcal{E}_{(s)} \leftrightarrow \mathcal{U}(H^{F^*})$, $\rho \leftrightarrow \rho_1$, such that*

$$\langle \rho, R_{T^*,s}^G \rangle = \varepsilon_{(s)} \langle \rho_1, R_{T^*,1}^H \rangle$$

for all F^*-stable maximal tori $T^* \subseteq G^*$ with $s \in T^*$.

This result is in fact a consequence of Theorem 9.3 and the Main Theorem 4.23 in Lusztig (1984a). The latter establishes a parametrization of $\mathcal{E}_{(s)}$ which only depends on the Weyl group of $C_{G^*}(s)$ and the action of F on it. Thus, it also gives a parametrization in terms of the unipotent characters of $C_{G^*}(s)$. The parametrization of unipotent characters will be explained in more detail in the following section.

There is a special case where the bijection in the above theorem is given by the operation of twisted induction. This is the case when $C_{G^*}(s)$ lies in a regular subgroup of G^*, see Lusztig (1976) and Chap. 13 in Digne and Michel (1991). If this condition is not satisfied, the element s is called *isolated.* It can be shown that if G is semisimple then there are only finitely many conjugacy classes of isolated semisimple elements (see, for example, Digne and Michel (1991), Lemma 14.11).

Lusztig (1988) extends the Jordan decomposition of characters to the case where the center of G is not necessarily connected. Let $s \in G^{*F^*}$ be semisimple and $H := C_{G^*}(s)$. Then H is not necessarily connected, and we let H° be its connected component. In this case, we define $\mathcal{U}(H^{F^*})$ to be the set of all irreducible characters of H^{F^*} whose restriction to $H^{\circ F^*}$ is a sum of unipotent characters. Furthermore, we define $R^H_{T^*,1}$ to be the generalized character of H^{F^*} obtained by inducing $R^{H^\circ}_{T^*,1}$ to H^{F^*}. With these definitions, the statement of Theorem 9.4 remains true, see also Digne and Michel (1990).

10 The multiplicity formula

The Jordan decomposition of characters shows the importance of unipotent characters. We will now explain the multiplicity formula of the Main Theorem 4.23 in Lusztig (1984a) for these characters. The problem we are concerned with is that of determining the scalar products

$$\langle \rho, R^G_{T,1} \rangle$$

where $\rho \in \mathcal{U}(G^F)$ and $T \subseteq G$ is an F-stable maximal torus.

Recall that the Frobenius map F induces an automorphism of (W, S) of finite order which we denote by the same symbol. Then this automorphism also acts on $\mathrm{Irr}(W)$, and we denote by $\mathrm{Irr}(W)^F$ the set of irreducible characters of W which are invariant under this action. Let $\tilde{W}$ be the semidirect product of W with the infinite cyclic group with generator γ, so that in $\tilde{W}$, we have the identity $\gamma \cdot w \cdot \gamma^{-1} = F(w)$ $(w \in W)$. If $\varphi \in \mathrm{Irr}(W)^F$ we can extend φ to an irreducible character $\tilde{\varphi}$ of $\tilde{W}$ which factors through a finite quotient of $\tilde{W}$. By Proposition 3.2 in Lusztig (1984a) there are precisely two such extensions which can be realized over $\mathbb{Q}$. We take such an extension $\tilde{\varphi}$ and define the class function

$$R_{\tilde{\varphi}} := \frac{1}{|W|} \sum_{w \in W} \tilde{\varphi}(\gamma \cdot w) R^G_{T_w,1}, \quad .$$

where, as before, $T_w \subseteq G$ is an F-stable maximal torus in the F-conjugacy class containing $w \in W$. Conversely, using orthogonality relations for the characters of $\tilde{W}$, we can also express the generalized characters $R_{T,1}^G$ as linear combinations of $R_{\tilde{\varphi}}$'s. Hence knowing the decomposition of the $R_{T,1}^G$'s into irreducible characters is equivalent to knowing the decomposition of the $R_{\tilde{\varphi}}$'s. Note that, a priori, $R_{\tilde{\varphi}}$ is a rational linear combination of irreducible characters of G^F.

Theorem 10.1 (Lusztig (1984a)) *Two characters $R_{\tilde{\varphi}}$ and $R_{\tilde{\varphi}'}$ as above have an irreducible constituent in common if and only if φ, φ' lie in the same family of* Irr(W).

Thus, we have a partition $\mathcal{U}(G^F) = \coprod_{\mathcal{F}} \mathcal{E}_{(1),\mathcal{F}}$, where $\mathcal{F}$ runs over the F-stable families in Irr(W). A unipotent character ρ lies in $\mathcal{E}_{(1),\mathcal{F}}$ if and only if ρ has non-zero scalar product with $R_{\tilde{\varphi}}$ for some $\varphi \in \mathcal{F}$.

The final step is to parametrize the characters in $\mathcal{E}_{(1),\mathcal{F}}$ for an F-stable family $\mathcal{F} \subset \mathrm{Irr}(W)$. Let $\tilde{\mathcal{F}}$ be the set of irreducible characters of $\tilde{W}$ which factor through a finite quotient and whose restrictions to W are irreducible and lie in $\mathcal{F}$. Following Chap. 4 in Lusztig (1984a), one can associate with $\mathcal{F}$ and the action of F on W two groups $\mathcal{G}_{\mathcal{F}} \subset \tilde{\mathcal{G}}_{\mathcal{F}}$ such that $\mathcal{G}_{\mathcal{F}}$ is finite and normal in $\tilde{\mathcal{G}}_{\mathcal{F}}$ and the quotient is isomorphic to $\mathbb{Z}$. This is done case by case, starting with irreducible finite Weyl groups and then taking into account direct products and automorphisms acting on them.

Let $\mathcal{G}_{\mathcal{F}}'$ be the preimage of $1 \in \mathbb{Z}$ under the canonical map $\tilde{\mathcal{G}}_{\mathcal{F}} \to \mathbb{Z}$. We define $\mathcal{M}(\mathcal{G}_{\mathcal{F}} \subset \tilde{\mathcal{G}}_{\mathcal{F}})$ to be the set of all pairs (y, τ) where y is an element of $\mathcal{G}_{\mathcal{F}}$ whose centralizer in $\tilde{\mathcal{G}}_{\mathcal{F}}$ meets $\mathcal{G}_{\mathcal{F}}'$ and τ is an irreducible character of that centralizer whose restriction to $C_{\mathcal{G}_{\mathcal{F}}}(y)$ is still irreducible and which factors through a finite quotient; these pairs are taken modulo the equivalence relation $(y, \tau) \sim (gyg^{-1}, \tau^g)$ $(g \in \tilde{\mathcal{G}}_{\mathcal{F}})$. We also define $\overline{\mathcal{M}}(\mathcal{G}_{\mathcal{F}} \subset \tilde{\mathcal{G}}_{\mathcal{F}})$ to be the set of all pairs $(x, \bar{\sigma})$ where x is an element of $\mathcal{G}_{\mathcal{F}}'$ and $\bar{\sigma}$ is an irreducible character of $C_{\mathcal{G}_{\mathcal{F}}}(x)$; these pairs are taken modulo an equivalence relation as above. There is a pairing $\{\,,\,\}\colon \overline{\mathcal{M}}(\mathcal{G}_{\mathcal{F}} \subset \tilde{\mathcal{G}}_{\mathcal{F}}) \times \mathcal{M}(\mathcal{G}_{\mathcal{F}} \subset \tilde{\mathcal{G}}_{\mathcal{F}}) \longrightarrow \bar{\mathbb{Q}}_\ell$ defined by

$$\{(x, \bar{\sigma}), (y, \tau)\} = \sum_g \frac{1}{|C_{\mathcal{G}_{\mathcal{F}}}(x)|} \frac{1}{|C_{\mathcal{G}_{\mathcal{F}}}(y)|} \tau(g^{-1}x^{-1}g)\bar{\sigma}(gyg^{-1})$$

where the sum is over all $g \in \mathcal{G}_{\mathcal{F}}$ such that $xgyg^{-1} = gyg^{-1}x$. Finally, Lusztig defines an embedding $\tilde{\mathcal{F}} \hookrightarrow \mathcal{M}(\mathcal{G}_{\mathcal{F}} \subset \tilde{\mathcal{G}}_{\mathcal{F}})$ and a function $\Delta\colon \overline{\mathcal{M}}(\mathcal{G}_{\mathcal{F}} \subset \tilde{\mathcal{G}}_{\mathcal{F}}) \to \{\pm 1\}$, again case by case. The function Δ is identically 1 except possibly if W has some factors of type E_7 or E_8. With this notation we can now state:

Theorem 10.2 (Lusztig (1984a)) *Let $\mathcal{F}$ be an F-stable family of* Irr(W). *Then there exists a bijection*

$$\mathcal{E}_{(1),\mathcal{F}} \leftrightarrow \overline{\mathcal{M}}(\mathcal{G}_{\mathcal{F}} \subset \tilde{\mathcal{G}}_{\mathcal{F}}), \quad \rho \leftrightarrow \bar{x}_\rho,$$

such that for any $\rho \in \mathcal{E}_{(1),\mathcal{F}}$ *and any* $\varphi \in \mathcal{F}$ *we have*

$$\langle \rho, R_{\tilde{\varphi}} \rangle = \Delta(\bar{x}_\rho)\{\bar{x}_\rho, x_{\tilde{\varphi}}\},$$

where $x_{\tilde{\varphi}}$ *corresponds to* $\tilde{\varphi}$ *under the embedding* $\tilde{\mathcal{F}} \hookrightarrow \mathcal{M}(\mathcal{G}_\mathcal{F} \subset \tilde{\mathcal{G}}_\mathcal{F})$.

Remark 10.3 The definition of the groups $\mathcal{G}_\mathcal{F}$, $\tilde{\mathcal{G}}_\mathcal{F}$, the corresponding sets $\mathcal{M}(\mathcal{G}_\mathcal{F} \subset \tilde{\mathcal{G}}_\mathcal{F})$, $\overline{\mathcal{M}}(\mathcal{G}_\mathcal{F} \subset \tilde{\mathcal{G}}_\mathcal{F})$, the pairing $\{\,,\}$ and the function Δ only depends on the finite Weyl group W, the action of F on it, and the family $\mathcal{F}$. In this sense, the classification of unipotent characters is "independent of q".

Remark 10.4 Let M be the group of all roots of unity in $\overline{\mathbb{Q}}_\ell$. Then there are natural free actions of M on $\tilde{\mathcal{F}}$ and $\mathcal{M}(\mathcal{G}_\mathcal{F} \subset \tilde{\mathcal{G}}_\mathcal{F})$, and the embedding $\tilde{\mathcal{F}} \hookrightarrow \mathcal{M}(\mathcal{G}_\mathcal{F} \subset \tilde{\mathcal{G}}_\mathcal{F})$ is M-equivariant. Moreover, the orbit space of $\mathcal{M}(\mathcal{G}_\mathcal{F} \subset \tilde{\mathcal{G}}_\mathcal{F})$ has the same cardinal as the set $\overline{\mathcal{M}}(\mathcal{G}_\mathcal{F} \subset \tilde{\mathcal{G}}_\mathcal{F})$. For each $x \in \mathcal{M}(\mathcal{G}_\mathcal{F} \subset \tilde{\mathcal{G}}_\mathcal{F})$ we define the corresponding *unipotent almost character* to be the class function

$$R_x := \sum_{\rho \in \mathcal{E}_{(1),\mathcal{F}}} \{\bar{x}_\rho, x\}\Delta(\bar{x}_\rho)\rho.$$

This reduces to $R_{\tilde{\varphi}}$ if x is the image of $\tilde{\varphi}$ under the above embedding. Up to a root of unity, R_x only depends on the M-orbit of x. Let

$$X_0 := \coprod_{\mathcal{F}} \mathcal{M}(\mathcal{G}_\mathcal{F} \subset \tilde{\mathcal{G}}_\mathcal{F})/M,$$

where $\mathcal{F}$ runs over all F-stable families in $\mathrm{Irr}(W)$. Then $\{R_x \mid x \in X_0\}$ and $\{\rho \mid \rho \in \mathcal{U}(G^F)\}$ generate the same subspace of the space of class functions on G^F, and each of these sets is an orthonormal basis for that subspace.

Remark 10.5 It is known that if G is of classical type then every $\rho \in \mathcal{U}(G^F)$ is uniquely determined by the multiplicities with which it occurs in the various characters $R_{\tilde{\varphi}}$. More generally, if G is arbitrary, a uniqueness statement is obtained by taking into account:

- the multiplicity formula in Theorem 10.2 above,
- the eigenvalues of Frobenius associated with a unipotent character (see (2.20), (3.8)(i) and Chap. 11 in Lusztig (1984a)),
- a compatibility with the parametrization of unipotent principal series characters by $\mathrm{Irr}(W^F)$.

For more details, see Part II in Digne and Michel (1990).

11 Computing character values

In this final section we briefly discuss the problem of actually computing character values, and hence of determining the complete character tables of finite groups of Lie type. (A more detailed account would go far beyond the scope of this survey article.)

We extend the definition of almost characters to all Lusztig series by using the Jordan decomposition of characters. For each semisimple $s \in G^{*F^*}$ we let X_0^s be the parameter set X_0 of Remark 10.4 but defined with respect to $C_{G^*}(s)$. Note that, a priori, this works only if the center of G is connected; in the general case, this has to be done in the framework of Lusztig (1988) or Digne and Michel (1990). Let $X_0^G = \coprod_{(s)} X_0^s$ where (s) runs over the F-stable classes of semisimple elements in G^*. Thus, $\{R_x^s \mid x \in X_0^G\}$ is a new orthonormal basis of the space of class functions on G^F. The transformation to the basis consisting of the irreducible characters of G^F is explicitly known. Thus, the problem of computing the values of the irreducible characters of G^F is equivalent to that of computing the values of the almost characters.

As far as the latter problem is concerned, the results are less complete than for the classification of the irreducible characters or the almost characters itself. The framework for attacking that problem is provided by Lusztig's theory of character sheaves, developed in Lusztig (1985/1986). Character sheaves are objects in the derived category of constructible $\bar{\mathbb{Q}}_\ell$-sheaves on G. If such an object is invariant under the Frobenius map F it gives rise to a so-called characteristic function which is a class function on G^F whose values can be explicitly computed. Under some conditions on p, q or on the center of G, these characteristic functions turn out to coincide with the almost characters of G^F, up to an algebraic number of absolute value 1, see Lusztig (1992) and Shoji (1995).

To illustrate these ideas, we shall present a special case of these results in a more precise form. For this purpose, we assume that G is simple of adjoint type. We construct a basis of the space of class functions on G^F as follows. Let C be an F-stable conjugacy class, $x_0 \in C^F$ and $A(x_0)$ the group of components of $C_G(x_0)$, as in Example 4.3. Then F induces an action on $A(x_0)$. We consider pairs $i = (C, \psi)$ where ψ is an F-invariant irreducible character of $A(x_0)$. For any such pair we let Y_i be the class function on G^F defined by $Y_i(g) = 0$ if $g \in G^F \setminus C^F$ and $Y_i(g_a) = \tilde{\psi}(Fa)$ where $\tilde{\psi}$ is an extension of ψ to the semidirect product of $A(x_0)$ with the cyclic subgroup of $\mathrm{Aut}(A(x_0))$ generated by F, and $g_a \in C^F$ lies in the G^F-class parametrized by the F-conjugacy class of $a \in A(x_0)$, cf. Section 4. (Note that Y_i depends on the choice of an extension $\tilde{\psi}$, and hence is only well-defined up to a non-zero scalar multiple.) The set of all Y_i where $i = (C, \psi)$ is a pair as above then forms the desired basis of the space of class functions on G^F.

Theorem 11.1 *Recall that G is simple of adjoint type. Assume that $p \neq 2$ if*

G has type E_6, $p \neq 2,3$ if G has type G_2, F_4, E_7, and $p \neq 2,3,5$ if G has type E_8. Let $s \in G^{*F^*}$ be semisimple and $x \in X_0^s$ be such that R_x^s is orthogonal to all class functions $R_{L,P}^G(f)$ where $L \subset G$ is a proper regular subgroup and f is any class function on L^F. Then there exists an F-stable pair $i = (C, \psi)$ as above and an algebraic number ζ_i of absolute value 1 such that

$$R_x^s = q^{(\dim G - \dim C)/2} \zeta_i Y_i.$$

The proof involves the following ingredients. By Shoji (1995), any almost character R_x^s coincides, up to a non-zero scalar multiple, with the characteristic function of some F-stable character sheaf A on G. By Shoji (1996a) (which extends earlier results of Lusztig (1990)), the induction of character sheaves from a regular subgroup of G coincides, on the level of characteristic functions, with the operation of twisted induction (cf. Section 7). Thus, the assumption on R_x^s in the above theorem means that the corresponding character sheaf is "cuspidal". By Theorem 23.1 in Lusztig (1985/1986), a cuspidal character sheaf is "clean". This means that its characteristic function coincides with some function Y_i as above, up to scalar multiple. The condition that almost characters have norm 1 then yields the above result.

As explained in Lusztig (1992), the above theorem – or extensions thereof for groups not necessarily simple of adjoint type – provides the basis for an inductive approach to the computation of the values of all almost characters, and hence of all irreducible characters of G^F. In order to carry out the induction, one also has to find a way to compute the values of non-cuspidal almost characters. It is in fact possible to write down an algorithm solving this problem, which involves the following ingredients:

- the generalized Springer correspondence, see Lusztig (1984b).
- an algorithm for the computation of generalized Green functions, see Shoji (1987) and Lusztig (1985/1986), Chap. 24.
- a character formula for the twisted induction of cuspidal almost characters, see Lusztig (1985/1986), Theorem 8.5, and Lusztig (1990).

The only problem left unsolved by this plan is that of calculating the numbers ζ_i in Theorem 11.1. In the case where G is of classical type and $s = 1$ this problem is solved in Shoji (1996b). In the case where the given cuspidal almost character is non-zero on some unipotent element, it is solved in Lusztig (1986).

References

Benson, C.T., Curtis, C.W. (1972)

'On the degrees and rationality of certain characters of finite Chevalley groups', *Trans. Amer. Math. Soc.* **165**, 251–273. See also the 'Corrections and additions', *ibid.* **202** (1975), 405–406.

Borel, A. (1991) 'Linear algebraic groups', *Graduate Texts in Mathematics* **126**, Springer.

Borel, A., et al. (1970)

'Seminar on algebraic groups and related finite groups', *Lecture Notes in Mathematics* **131**, Springer.

Carter, R.W. (1985)

'Finite groups of Lie type: Conjugacy classes and complex characters', Wiley, New York.

Chevalley, C. (1955)

'Sur certains groupes simples', *Tôhoku Math. J.* **7**, 14–66.

Deligne, P., Lusztig, G. (1976)

'Representations of reductive groups over finite fields', *Annals Math.* **103**, 103–161.

Digne, F., Michel, J. (1990)

'On Lusztig's parametrization of characters of finite groups of Lie type', *Astérisque* **181–182**, 113–156.

Digne, F., Michel, J. (1991)

'Representations of finite groups of Lie type', *London Math. Soc. Students Texts* **21**, Cambridge University Press.

Geck, M., Hiss, G. (1996)

'Modular representations of finite groups of Lie type in non-defining characteristic', *Progress in Math.* **141**, 195–249, Birkhäuser, Boston.

Lusztig, G. (1976)

'On the finiteness of the number of unipotent classes', *Invent. Math.* **34**, 201–213.

Lusztig, G. (1977)

'Irreducible representations of finite classical groups', *Invent. Math.* **43**, 125–175.

Lusztig, G. (1984a)

'Characters of reductive groups over a finite field', *Annals Math. Studies* **107**, Princeton University Press.

Lusztig, G. (1984b)

'Intersection cohomology complexes on a reductive group', *Invent. Math.* **75**, 205–272.

Lusztig, G. (1985/1986)

'Character sheaves', *Advances in Math.* **56**, 193–237; II, **57**, 226–265; III, **57**, 266–315 (1985); IV, **59**, 1–63; V, **61**, 103–155.

Lusztig, G. (1986)

'On the character values of finite Chevalley groups at unipotent elements', *J. Algebra* **104**, 146–194.

Lusztig, G. (1987)

'Leading coefficients of character values of Hecke algebras', *Proc. Symp. in Pure Math.* **47**, 235–262.

Lusztig, G. (1988)

'On the representations of reductive groups with disconnected centre', *Astérisque* **168**, 157–166.

Lusztig, G. (1990)

'Green functions and character sheaves', *Annals of Math.* **131**, 355–408.

Lusztig, G. (1992)

'Remarks on computing irreducible characters', *J. Amer. Math. Soc.* **5**, 971–986.

Milne, J.S. (1980)

'Étale cohomology', *Princeton University Press.*

Shoji, T. (1987)

'Green functions of reductive groups over a finite field,' *Proc. Symp. in Pure Math.* **47**, 289–302.

Shoji, T. (1995)

'Character sheaves and almost characters of reductive groups', *Advances in Math.* **111**, 244–313; II, *ibid.*, 314–354.

Shoji, T. (1996a)

'On the computation of unipotent characters of finite classical groups', *AAECC* **7**, 165–174.

Shoji, T. (1996b)

'Unipotent characters of finite classical groups', *Progress in Math.* **141**, 373–414, Birkhäuser, Boston.

Srinivasan, B. (1979)

'Representations of finite Chevalley groups', *Lecture Notes in Mathematics* **764**, Springer.

Steinberg, R. (1968)

'Endomorphisms of linear algebraic groups', *Memoirs Amer. Math. Soc.* **80**.

Steinberg, R. (1974)

'Conjugacy classes in algebraic groups', *Lecture Notes in Mathematics* **366**, Springer.

Generalized Harish-Chandra Theory

Michel Broué & Gunter Malle

Université Paris 7 Denis–Diderot et Institut Universitaire de France, Case 7012, 2 Place Jussieu, F–75251 Paris Cedex 05, France

IWR, Universität Heidelberg, 69120 Heidelberg, Germany

1 Introduction

In the study of subgroup structure, ordinary and modular representation theory of finite groups of Lie type it turns out that many properties behave in a generic manner, i.e., the results can be phrased in terms independent of the order q of the field of definition. It is the aim of this article to present a framework in which this generic behaviour can be conveniently formulated (and partly also proved).

More precisely, assume we are given a family of groups of Lie type, like the groups $\mathrm{GL}_n(q)$ for fixed n but varying prime power q. Then a crucial role in the description of this family of groups is played by the Weyl group, which is the same for all members in the family, together with the action induced by the Frobenius morphism (the twisting). The Weyl group occurs in a natural way as a reflection group on the vector space generated by the coroots, and the Frobenius morphism induces an automorphism of finite order. This leads to the concept of a generic finite reductive group $\mathbb{G}$. The orders of the groups attached to a generic finite reductive group $\mathbb{G}$ can be obtained as values of one single polynomial, the order polynomial $|\mathbb{G}|$ of $\mathbb{G}$. In analogy to the concept of ℓ-subgroup of a finite group, for any cyclotomic polynomial Φ_d dividing the order polynomial $|\mathbb{G}|$ one can define d-tori of $\mathbb{G}$. These satisfy a complete analogue of the Sylow theorems for finite groups. These concepts were first formalized in [1].

In the representation theory of groups of Lie type, the right generic objects are the unipotent characters. It follows from results of Lusztig that they can be parametrized in terms only depending on the generic group $\mathbb{G}$. Moreover, the functor of twisted induction can be shown to be generic. It turns out that for any d such that Φ_d divides $|\mathbb{G}|$ there holds a d-Harish-Chandra theory for unipotent characters. This gives rise to a whole family of generalized Harish-Chandra theories which contains the usual Harish-Chandra theory for complex characters as the case $d = 1$. Surprisingly, the resulting generalized

Harish-Chandra series allow a description of the subdivision into ℓ-blocks of the unipotent characters of a finite reductive group, at least for large primes ℓ, thus giving an application of the generic formalism to the ℓ-modular representation theory of groups of Lie type. The generalized Harish-Chandra series can be indexed by irreducible characters of suitable relative Weyl groups such that the decomposition of the functor of twisted induction can be compared to ordinary induction in these relative Weyl groups. This was proved in [3]. These relative Weyl groups turn out to be pseudo-reflection groups.

We introduce generic groups and their generic order in Sections 2 and 3 and formulate the d-Sylow theorems in Section 4. In Section 5 the setup and the main statements of ordinary Harish-Chandra theory are recalled. In Sections 6 and 7 we introduce generic unipotent characters and describe the generalized Harish-Chandra theories. The application to ℓ-blocks is given in Section 8. In the last section we discuss the decomposition of twisted induction and the role played in this by the relative Weyl group and (conjecturally) by its cyclotomic Hecke algebra.

2 Generic finite reductive groups

2.1 From algebraic groups to generic groups

We start with a connected reductive algebraic group $\mathbf{G}$ over the algebraic closure of a finite field of characteristic $p > 0$. We assume that $\mathbf{G}$ is already defined over a finite field and let $F : \mathbf{G} \to \mathbf{G}$ be the corresponding Frobenius morphism. The group of fixed points $\mathbf{G}^F$ is then a finite group of Lie type. Recall from Geck's lecture that the choice of an F-stable maximal torus $\mathbf{T}$ of $\mathbf{G}$ and a Borel subgroup containing $\mathbf{T}$ gives rise to a root datum $(X, R, Y, R^\vee)$, consisting of the character and cocharacter groups X, Y of $\mathbf{T}$, the set of roots $R \subset X$ and the set of coroots $R^\vee \subset Y$. The Frobenius map F acts on $Y_\mathbb{R} := Y \otimes_\mathbb{Z} \mathbb{R}$ as $q\phi$ where q is a power of p and ϕ is an automorphism of finite order. Replacing the Borel subgroup by another one containing $\mathbf{T}$ changes ϕ by an element of the Weyl group W of $\mathbf{G}$ with respect to $\mathbf{T}$. Hence ϕ is uniquely determined as automorphism of $Y_\mathbb{R}$ up to elements of W.

We can thus naturally associate to $(\mathbf{G}, \mathbf{T}, F)$ the data $(X, R, Y, R^\vee, W\phi)$. Here,

(i) X, Y are free $\mathbb{Z}$-modules of equal finite rank, endowed with a duality $X \times Y \to \mathbb{Z}$, $(x, y) \mapsto \langle x, y \rangle$,

(ii) $R \subset X$ and $R^\vee \subset Y$ are root systems with a bijection $R \to R^\vee$, $\alpha \mapsto \alpha^\vee$, such that $\langle \alpha, \alpha^\vee \rangle = 2$ (see Section 6 in Rouquier's lecture),

(iii) W is the Weyl group of the root system $R^\vee$ in Y and ϕ is an automorphism of Y of finite order stabilizing $R^\vee$.

A quintuple $(X, R, Y, R^\vee, W\phi)$ satisfying these properties is called a *generic finite reductive group* or *complete root datum.*

Conversely, let's start from a complete root datum $\mathbb{G} = (X, R, Y, R^\vee, W\phi)$. Then for any choice of a prime number p, $\mathbb{G}$ determines a pair $(\mathbf{G}, \mathbf{T})$ as above, up to inner automorphisms of $\mathbf{G}$ induced by $\mathbf{T}$ (see the lecture of Carter). Moreover, the additional choice of a power q of p determines a triple $(\mathbf{G}, \mathbf{T}, F)$ as above. In this way the generic finite reductive group $\mathbb{G}$ gives rise to a whole series $\{\mathbb{G}(q) := \mathbf{G}^F \mid q \text{ a prime power}\}$ of groups of Lie type.

Example 2.1 Let $\mathbf{G} = \mathrm{GL}_n(\overline{\mathbb{F}_q})$ be the group of invertible $n \times n$-matrices over the algebraic closure of the finite field $\mathbb{F}_q$ with the maximal torus $\mathbf{T}$ consisting of the diagonal matrices in $\mathbf{G}$. Then $\mathbf{T}$ is F-stable for the Frobenius map $F : \mathbf{G} \to \mathbf{G}$ which raises every matrix entry to its qth power, as well as for the product F^- of F with the transpose-inverse map on $\mathbf{G}$. In the first case, the group of F-fixed points is the general linear group over $\mathbb{F}_q$, while for the second Frobenius map F^- we obtain the general unitary group $\mathrm{GU}_n(q)$. One easily checks that $(\mathbf{G}, \mathbf{T}, F)$ gives rise to a generic finite reductive group of the form

$$\mathrm{GL}_n = (\mathbb{Z}^n, R, \mathbb{Z}^n, R^\vee, \mathfrak{S}_n \cdot \mathrm{Id})$$

with $R = R^\vee = \{e_i - e_j \mid i \neq j\}$, where $\{e_1, \dots, e_n\}$ is the standard basis of $\mathbb{Z}^n$, while $(\mathbf{G}, \mathbf{T}, F^-)$ gives rise to $\mathrm{GU}_n = (\mathbb{Z}^n, R, \mathbb{Z}^n, R^\vee, \mathfrak{S}_n(-\mathrm{Id}))$. In this sense we may think of $\mathrm{GU}_n(q)$ as being $\mathrm{GL}_n(-q)$.

2.2 Generic Levi subgroups and tori

A *generic Levi subgroup* of $\mathbb{G} = (X, R, Y, R^\vee, W_{\mathbb{G}}\phi)$ is by definition a generic finite reductive group of the form

$$\mathbb{L} = (X, R', Y, R'^\vee, W_{\mathbb{L}} w\phi),$$

where $w \in W_{\mathbb{G}}$ and $R'^\vee$ is a $w\phi$-stable parabolic subsystem of $R^\vee$ with Weyl group $W_{\mathbb{L}}$. A *generic torus* is a generic finite reductive group with $R = R^\vee = \emptyset$. Thus a generic torus of $\mathbb{G}$ has the form $(X', Y', (w\phi)|_{Y'})$ for some $w \in W_{\mathbb{G}}$ and some $w\phi$-stable direct summand Y' of Y, where X' is the dual of Y', i.e., $X' = X/(Y')^\perp$ with

$$(Y')^\perp = \{x \in X \mid \langle x, y \rangle = 0 \text{ for all } y \in Y'\}.$$

The Weyl group $W_{\mathbb{G}}$ acts naturally on the set of generic Levi subgroups of $\mathbb{G}$ as well as on the set of generic tori of $\mathbb{G}$. The following result allows to switch freely between the language of generic finite reductive groups and that of actual groups of Lie type (see [1, Th. 2.1]):

Proposition 2.2 *Let $\mathbb{G}$ be a generic finite reductive group and q a prime power, and let $(\mathbf{G}, \mathbf{T}, F)$ be the associated finite reductive group.*

(a) There exists a natural bijection between the $W_{\mathbb{G}}$-classes of generic Levi subgroups of $\mathbb{G}$ and the $\mathbf{G}^F$-classes of F-stable Levi subgroups of $\mathbf{G}$.

(b) There exists a natural bijection between the $W_{\mathbb{G}}$-classes of generic tori of $\mathbb{G}$ and the $\mathbf{G}^F$-classes of F-stable tori of $\mathbf{G}$

We sketch the construction of the bijection in (a). Let $\mathbf{S} \leq \mathbf{G}$ be an F-stable torus in $\mathbf{G}$. Then $\mathbf{S}$ is contained in some F-stable maximal torus of $\mathbf{G}$. Since all maximal tori of $\mathbf{G}$ are conjugate, there exists a $g \in \mathbf{G}$ with $\mathbf{S} \leq {}^g\mathbf{T}$. Then $g^{-1}F(g)$ normalizes $\mathbf{T}$, and we denote its image in the Weyl group $W = N_{\mathbf{G}}(\mathbf{T})/\mathbf{T}$ by w. The cocharacter group Y' of $\mathbf{S}^g$ can be identified with a subgroup of the cocharacter group Y of $\mathbf{T}$, and similarly the character group X' of $\mathbf{S}^g$ with a quotient of X. We now associate to $\mathbf{S}$ the generic torus $\mathbb{S} := (X', Y', (w\phi)|_{Y'})$ of $\mathbb{G}$. Note that by construction Y' is $w\phi$-stable. Conversely, if $\mathbb{S} = (X', Y', (w\phi)|_{Y'})$ is a torus of $\mathbb{G}$, choose $n \in N_{\mathbf{G}}(\mathbf{T})$ with image w. The inclusion $Y' \leq Y$ defines an nF-stable subtorus $\mathbf{S}_w$ of $\mathbf{T}$. By the theorem of Lang (see Theorem 3.5 in Geck's lecture) there exists a $g \in \mathbf{G}$ with $n = g^{-1}F(g)$. Then $\mathbf{S} := {}^g\mathbf{S}_w$ is an F-stable torus of $\mathbf{G}$.

Part (b) is essentially a consequence of the following connection between the concepts of generic tori and generic Levi subgroups. For a generic torus $\mathbb{S} = (X', Y', (w\phi)|_{Y'})$ of $\mathbb{G}$ we define its *centralizer* to be the generic Levi subgroup

$$C_{\mathbb{G}}(\mathbb{S}) := (X, R', Y, R'^{\vee}, W'w\phi),$$

where R' consists of those roots $\alpha \in R$ which are orthogonal to Y', and W' is the Weyl group of the root system $R'^{\vee}$. For a generic Levi subgroup $\mathbb{L} = (X, R', Y, R'^{\vee}, W_{\mathbb{L}}w\phi)$ we define its *center* to be the generic torus

$$Z(\mathbb{L}) := (X', Y', (w\phi)|_{Y'})$$

of $\mathbb{G}$, where Y' is the orthogonal of R' in Y. With these definitions it is easy to see that any generic Levi subgroup $\mathbb{L}$ of $\mathbb{G}$ satisfies $C_{\mathbb{G}}(Z(\mathbb{L})) = \mathbb{L}$ (see [1, Prop. 1.3]). Furthermore, if $\mathbf{S}$ is an F-stable torus of the algebraic group $\mathbf{G}$ and $\mathbf{L}$ is its centralizer in $\mathbf{G}$, then the generic group $\mathbb{L}$ of $\mathbf{L}$ is the centralizer in $\mathbb{G}$ in the sense defined before of the generic group $\mathbb{S}$ of $\mathbf{S}$, and similarly for the center of a Levi.

2.3 Some constructions

Many constructions for algebraic groups have a convenient formulation in terms of generic groups. Let $\mathbb{G} = (X, R, Y, R^{\vee}, W\phi)$ be a generic group. The dual generic group (see Section 8 in the lecture of Geck) is defined by $\mathbb{G}^* = (Y, R^{\vee}, X, R, W\phi^{\vee -1})$, where $\phi^{\vee}$ is the automorphism of X adjoint to ϕ. We write $Q(R)$ for the $\mathbb{Z}$-submodule of X generated by R. Then the *radical of* $\mathbb{G}$ (see Section 2 in the lecture of Carter) is the generic torus

$$\mathrm{Rad}(\mathbb{G}) := (X/Q(R)^{\perp\perp}, Q(R)^{\perp}, \phi|_{Q(R)^{\perp}}),$$

the *semisimple quotient of* $\mathbb{G}$ is

$$\mathbb{G}_{ss} := (Q(R)^{\perp\perp}, R, Y/Q(R)^{\perp}, R^{\vee}, W\phi).$$

The reader is invited to verify in each case that our construction does indeed yield the generic group associated to the result of the corresponding operation on algebraic groups. Finally, for a positive integer d we denote by $\mathbb{G}^{(d)}$ the generic group $(X^d, R^d, Y^d, (R^{\vee})^d, W^d\sigma\phi)$, where $(X^d, R^d, Y^d, (R^{\vee})^d)$ is the direct product of d copies of the root datum of $\mathbb{G}$, and where σ is the automorphism of Y^d which permutes the d factors cyclically. One checks that $\mathbb{G}^{(d)}(q) \cong \mathbb{G}(q^d)$.

3 The polynomial order

We want to formulate a Sylow theory for generic groups. For this, we first have to say what should be the order $|\mathbb{G}|$ of a generic group $\mathbb{G}$, and then define what Φ-subgroups for prime divisors Φ of $|\mathbb{G}|$ are.

3.1 The order polynomial

Let $\mathbb{G} = (X, R, Y, R^{\vee}, W\phi)$ be a generic finite reductive group. The Weyl group W acts on $V := Y \otimes_{\mathbb{Z}} \mathbb{C}$ as a group generated by reflections. Thus its ring of invariants $S(V)^W$ in the symmetric algebra $S(V)$ is a polynomial ring by the theorem of Shephard/Todd and Chevalley (see Theorem 9.1 in Rouquier's lecture). Since the element ϕ acts completely reducibly on V it is possible to choose $n = \dim(V)$ algebraically independent homogeneous invariants $f_1, \ldots, f_n \in S(V)^W$ generating $S(V)^W$ which are eigenvectors for the action of ϕ. Thus $\phi f_i = \zeta_i f_i$ for some root of unity ζ_i. Denote the degree of f_i by d_i. Then it is known that the family $\{(d_1, \zeta_1), \ldots, (d_n, \zeta_n)\}$ is independent of the choice of the f_i (see for example [12, 6.1]). Let

$$\epsilon_{\mathbb{G}} := (-1)^n \zeta_1 \ldots \zeta_n = \det(-\phi) \in \{\pm 1\}$$

and let x be an indeterminate. We define the *polynomial order of* $\mathbb{G}$ to be

$$|\mathbb{G}| := \frac{\epsilon_{\mathbb{G}} x^N}{\frac{1}{|W|}\sum_{w\in W} \frac{1}{\det_V(1 - xw\phi)}},$$

where $N = |R|/2$ is the number of reflections in W. For example, if $\mathbb{G}$ is a torus, i.e., if $R = R^{\vee} = \emptyset$, $W = 1$, then the above formula becomes $|\mathbb{G}| = \epsilon_{\mathbb{G}} \det_V(1 - x\phi)$, hence the polynomial order of a torus (X, Y, ϕ) is just the characteristic polynomial of ϕ on V.

Proposition 3.1 *We have* $|\mathbb{G}| = x^N \prod_{i=1}^n (x^{d_i} - \zeta_i) \in \mathbb{Z}[x]$.

In the case $\phi = 1$ this is just Molien's formula for the ring of invariants $S(V)^W$. In the general case it follows by evaluating the graded trace of elements $w\phi$ on the graded $W\langle\phi\rangle$-module $S(V)$ in two ways.

It can be shown that for any choice of a prime power q the order of the corresponding finite group $\mathbb{G}(q)$ is given by

$$|\mathbb{G}(q)| = |\mathbb{G}|(q)$$

(see for example [5, Th. 9.4.10 and 14.3.1]). It is clear from Proposition 3.1 that $|\mathbb{G}|$ is a product of cyclotomic polynomials, that is, there exist non-negative integers $a(d)$ such that

$$|\mathbb{G}|(x) = x^N \prod_d \Phi_d(x)^{a(d)} ,$$

where $\Phi_d(x)$ denotes the dth cyclotomic polynomial over $\mathbb{Q}$, i.e., the polynomial whose roots are the primitive dth roots of unity. These cyclotomic polynomials will be considered to be the primes dividing the order of $\mathbb{G}$.

Example 3.2 We continue Example 2.1. For the generic group GL_n the order is given by

$$|\mathrm{GL}_n| = \prod_{i=0}^{n-1}(x^n - x^i) = x^{n(n-1)/2} \prod_d \Phi_d^{[\frac{n}{d}]}$$

where $[a]$ denotes the largest integer smaller than or equal to a. The formula for the polynomial order immediately shows that

$$|\mathrm{GU}_n|(x) = \pm|\mathrm{GL}_n|(-x).$$

This is an instance of the so-called Ennola-duality between GL_n and GU_n.

3.2 d-tori and d-split Levi subgroups

A generic torus $\mathbb{S}$ is called a *d-torus* if the order polynomial $|\mathbb{S}|(x)$ is a power of $\Phi_d(x)$. Thus by our above remark $\mathbb{S} = (X, Y, \phi)$ is a d-torus if and only if all eigenvalues of ϕ on V are primitive dth roots of unity. Translated to algebraic groups, a torus $\mathbb{S}$ is d-split if and only if the corresponding finite reductive groups $\mathbb{S}(q)$ are the groups of F-fixed points of an algebraic torus $\mathbf{S}$ which splits over $\mathbb{F}_{q^d}$ but no subtorus of which splits over any proper subfield.

The centralizers of generic d-tori of $\mathbb{G}$ are called *d-split Levi subgroups.* Note that, in particular, $\mathbb{G}$ itself is d-split. Thus any d such that Φ_d divides $|\mathbb{G}|$ gives rise to two families of substructures of $\mathbb{G}$: d-tori and d-split Levi subgroups.

Example 3.3 (a) Let $\mathbb{G}$ be a generic finite reductive group and $(\mathbf{G}, F)$ a finite reductive group determined from $\mathbb{G}$ and a choice of a prime power q. Then the 1-split Levis of $\mathbb{G}$ are precisely those generic Levi subgroups $\mathbb{L}$ of

$\mathbb{G}$ such that the corresponding F-stable Levi subgroup $\mathbf{L}$ of $\mathbf{G}$ is contained in an F-stable parabolic subgroup of $\mathbf{G}$. Thus the case $d = 1$ describes what is usually called the split or Harish-Chandra Levi subgroups. Similarly, the 1-tori are those generic tori $\mathbb{S}$ such that the corresponding F-stable torus $\mathbf{S}$ of $\mathbf{G}$ is split in the usual sense.

(b) We continue Example 2.1. If $(X', Y', w|_{Y'})$ is a d-torus of GL_n for $d = n$ then at least one eigenvalue of w on V has to be a primitive nth root of unity. The only elements with this property of $W = \mathfrak{S}_n$ in its permutation representation are the n-cycles (i.e., the Coxeter elements). Thus any non-trivial n-torus of GL_n is conjugate to $\mathbb{S} = (X', Y', w|_{Y'})$ where w is an n-cycle and Y' is the maximal sublattice of $Y = \mathbb{Z}^n$ such that all eigenvalues of w on $Y' \otimes_{\mathbb{Z}} \mathbb{C}$ are primitive n-th roots of unity (so Y' has rank $\varphi(n)$). The centralizer of $\mathbb{S}$ is the Coxeter torus (X, Y, w), which according to our above definition is just the generic group $\mathrm{GL}_1^{(n)}$. More generally, it is easily seen by considering cycle shapes of elements in $\mathfrak{S}_n$ that for arbitrary d the d-split Levi subgroups of GL_n have the type

$$\mathrm{GL}_{n_1}^{(d)} \times \ldots \times \mathrm{GL}_{n_r}^{(d)} \times \mathrm{GL}_s$$

for some $n_1, \ldots, n_r, s$ with $d(n_1 + \ldots + n_r) + s = n$.

The finite reductive groups associated to d-tori have a very simple structure (see [1, Prop. 3.3]):

Proposition 3.4 *Let $\mathbb{S}$ be a generic d-torus of order $\Phi_d^{a(d)}$. Let $\mathbb{S}(q)$ be the finite reductive group associated to $\mathbb{S}$ by the choice of a prime power q. Then $\mathbb{S}(q)$ is the direct product of $a(d)$ cyclic groups of order $\Phi_d(q)$.*

4 d-Sylow theorems

The d-tori introduced in the previous section behave like ℓ-subgroups of finite groups. To make this more precise we need one more definition. For a Levi subgroup $\mathbb{L}$ of a generic group $\mathbb{G}$ we define the *relative Weyl group of* $\mathbb{L}$ *in* $\mathbb{G}$ as

$$W_{\mathbb{G}}(\mathbb{L}) := N_{W_{\mathbb{G}}}(W_{\mathbb{L}})/W_{\mathbb{L}}\,.$$

Theorem 4.1 *Let $\mathbb{G}$ be a generic finite reductive group and d such that $\Phi_d(x)$ divides $|\mathbb{G}|$.*

(a) There exist non-trivial d-tori of $\mathbb{G}$.

(b) For any maximal d-torus $\mathbb{S}$ of $\mathbb{G}$ we have $|\mathbb{S}| = \Phi_d^{a(d)}$, where $a(d)$ is the precise power of Φ_d dividing $|\mathbb{G}|$.

(c) Any two maximal d-tori of $\mathbb{G}$ are conjugate.

(d) Let $\mathbb{S}$ be a maximal d-torus of $\mathbb{G}$ and $\mathbb{L} = C_{\mathbb{G}}(\mathbb{S})$ its centralizer. Then

$$\frac{|\mathbb{G}|}{|W_{\mathbb{G}}(\mathbb{L})||\mathbb{L}|} \equiv 1 \pmod{\Phi_d} \quad \textit{in } \mathbb{Z}[x].$$

Replacing 'generic finite reductive group' by 'finite group', 'd-torus' by 'ℓ-subgroup' and Φ_d by ℓ the first three statements of the theorem just become the familiar Sylow theorems for finite groups. In view of this, the maximal d-tori of $\mathbb{G}$ are called *generic Sylow d-tori.* The fourth assertion is the translation of the fact that the number of Sylow ℓ-subgroups of a finite group is congruent to 1 modulo ℓ.

We sketch the proof of the first three parts of this theorem. Since the order of a torus (X, Y, ϕ) is just the characteristic polynomial of ϕ on $V = Y \otimes_{\mathbb{Z}} \mathbb{C}$ we may translate assertions (a) to (c) to assertions about elements of $W_{\mathbb{G}}$ as follows: for $w \in W_{\mathbb{G}}$ and $\zeta \in \mathbb{C}$ let

$$V(w\phi, \zeta) := \{v \in V \mid w\phi v = \zeta v\}$$

be the eigenspace of $w\phi$ on V for the eigenvalue ζ. Then (a) states that for any d with $a(d) \neq 0$ there exists $w \in W_{\mathbb{G}}$ with $V(w\phi, \zeta_d) \neq 0$ where ζ_d is a primitive dth root of unity. The second part claims that there exists w such that $\dim(V(w\phi, \zeta_d)) = a(d)$, and the third part claims that all these maximal eigenspaces for a fixed ζ_d are conjugate under $W_{\mathbb{G}}$. It was shown by Springer [12, 3.4 and 6.2] that these three statements do hold for arbitrary finite groups generated by reflections, hence in particular for Weyl groups. The proof of the last part is of a somewhat different flavour (see [1, Th. 3.4]) and we will not go into this here.

Example 4.2 Let $\mathbb{G} = (X, R, Y, R^\vee, W\phi)$ be the generic finite reductive group attached to the series of Steinberg triality groups ${}^3D_4(q)$. Thus W is the Weyl group of type D_4 and ϕ is a non-trivial automorphism of W of order 3 induced by the inclusion of W into the Weyl group of type F_4. The Weyl group of type D_4 has polynomial invariants generated by homogeneous elements of degrees $2, 4, 6, 4$. The corresponding eigenvalues of ϕ are given by $1, \zeta_3, 1, \zeta_3^2$, with ζ_3 a primitive third root of unity. Thus $\mathbb{G}$ has polynomial order

$$|\mathbb{G}| = x^{12}(x^2-1)(x^8+x^4+1)(x^6-1) = x^{12}\Phi_1^2\Phi_2^2\Phi_3^2\Phi_6^2\Phi_{12}\,.$$

According to the Sylow theorems there exist (maximal) tori of $\mathbb{G}$ of orders $\Phi_3^2, \Phi_6^2, \Phi_{12}$, namely the Sylow d-tori for $d = 3, 6, 12$. Furthermore, there exist Sylow tori of orders Φ_1^2, Φ_2^2, but these tori are not maximal.

Let $\mathbb{G}$ be a generic group and q a prime power. The Sylow Theorem 4.1 also translates to an assertion about Sylow ℓ-subgroups of $\mathbb{G}(q)$ for large primes ℓ not dividing q (see [1, Cor. 3.13]):

Proposition 4.3 *Let ℓ be a prime dividing $|\mathbb{G}(q)|$ but not dividing $q|W\langle\phi\rangle|$.*

(a) There exists a unique d such that $\ell|\Phi_d(q)$ and $\Phi_d\big|\,|\mathbb{G}|$.

(b) Any Sylow ℓ-subgroup of $\mathbb{G}(q)$ is contained in the group $\mathbb{S}(q)$ for some Sylow d-torus $\mathbb{S}$ of $\mathbb{G}$.

(c) The Sylow ℓ-subgroups are isomorphic to a direct product of $a(d)$ cyclic groups of order ℓ^a, where $\Phi_d^{a(d)}$ is the precise power of Φ_d dividing $|\mathbb{G}|$ and where ℓ^a is the precise power of ℓ dividing $\Phi_d(q)$.

The first part is an easy exercise, while the second and third follow from Proposition 3.4 and Theorem 4.1.

5 Ordinary Harish-Chandra theory

We leave the generic point of view for a moment in order to give a brief account of Harish-Chandra theory for complex characters of finite groups of Lie type. In the next section we will return to the generic setting and explain how this ordinary Harish-Chandra theory can be considered as a special case of a whole family of such theories.

Let $\mathbf{G}$ be a connected reductive algebraic group defined over a finite field and let $F : \mathbf{G} \to \mathbf{G}$ be the corresponding Frobenius morphism. Let $\mathbf{P}$ be an F-stable parabolic subgroup of $\mathbf{G}$, with F-stable Levi-decomposition $\mathbf{P} = \mathbf{UL}$, i.e., $\mathbf{U}$ is the unipotent radical of $\mathbf{P}$ and $\mathbf{L}$ is an F-stable Levi complement. Any $\mathbb{C}\mathbf{L}^F$-module M can be considered as a $\mathbb{C}\mathbf{P}^F$-module via the projection map $\mathbf{P}^F \to \mathbf{L}^F$. The induced module of M from $\mathbf{P}^F$ to $\mathbf{G}^F$ is denoted by $R_{\mathbf{L},\mathbf{P}}^{\mathbf{G}}(M)$ (see Remark 7.2 in Geck's lecture). Conversely, for a $\mathbb{C}\mathbf{G}^F$-module M we denote by ${}^*R_{\mathbf{L},\mathbf{P}}^{\mathbf{G}}(M)$ the $\mathbb{C}\mathbf{L}^F$-module of $\mathbf{U}^F$-fixed points of M. It was first shown by Deligne that for fixed $\mathbf{L}$ the result of both constructions is independent of the choice of parabolic subgroup $\mathbf{P}$ containing $\mathbf{L}$. Let $\mathcal{L}$ be the set of F-stable Levi subgroups of F-stable parabolic subgroups of $\mathbf{G}$. Thus for any $\mathbf{L} \in \mathcal{L}$ we have linear maps

$$R_{\mathbf{L}}^{\mathbf{G}} : \mathbb{Z}\mathrm{Irr}(\mathbf{L}^F) \to \mathbb{Z}\mathrm{Irr}(\mathbf{G}^F), \quad {}^*R_{\mathbf{L}}^{\mathbf{G}} : \mathbb{Z}\mathrm{Irr}(\mathbf{G}^F) \to \mathbb{Z}\mathrm{Irr}(\mathbf{L}^F),$$

called *Harish-Chandra induction* and *restriction.* These are adjoint to each other with respect to the scalar product of characters. Moreover, Harish-Chandra induction is transitive in the sense that $R_{\mathbf{M}}^{\mathbf{G}} \circ R_{\mathbf{L}}^{\mathbf{M}} = R_{\mathbf{L}}^{\mathbf{G}}$ for $\mathbf{L}, \mathbf{M} \in \mathcal{L}$ with $\mathbf{L} \subset \mathbf{M}$ (see for example [6, Prop. 4.7]) and it satisfies a Mackey-formula (see [6, Th. 5.1]).

A key notion in Harish-Chandra theory is that of cuspidality. An irreducible character $\chi \in \mathrm{Irr}(\mathbf{G}^F)$ is called *cuspidal* if ${}^*R_{\mathbf{L}}^{\mathbf{G}}(\chi) = 0$ for all proper Levi subgroups $\mathbf{L} \in \mathcal{L}$. We then also say that the pair $(\mathbf{G}, \chi)$ is cuspidal. Let

$$\mathcal{K} := \{(\mathbf{L}, \lambda) \mid \mathbf{L} \in \mathcal{L},\ \lambda \in \mathrm{Irr}(\mathbf{L}^F)\}\,.$$

For $(\mathbf{L}_1, \lambda_1), (\mathbf{L}_2, \lambda_2) \in \mathcal{K}$ we write $(\mathbf{L}_1, \lambda_1) \leq (\mathbf{L}_2, \lambda_2)$ if $\mathbf{L}_1 \subset \mathbf{L}_2$ and λ_2 is a constituent of $R_{\mathbf{L}_1}^{\mathbf{L}_2}(\lambda_1)$. This defines a partial order relation on the set of pairs $\mathcal{K}$. By the adjointness of $R_{\mathbf{L}}^{\mathbf{G}}$ and ${}^*R_{\mathbf{L}}^{\mathbf{G}}$ the minimal elements are just the cuspidal pairs. The following Harish-Chandra principle is proved for example in [6, Sec. 6]:

Theorem 5.1 *(a) (Disjointness) For each $\gamma \in \mathrm{Irr}(\mathbf{G}^F)$ there exists a unique cuspidal pair $(\mathbf{L}, \lambda)$ (up to $\mathbf{G}^F$-conjugacy) such that γ occurs in $R_{\mathbf{L}}^{\mathbf{G}}(\lambda)$.*

(b) (Transitivity) Let $(\mathbf{L}, \lambda)$ be cuspidal and $(\mathbf{L}, \lambda) \leq (\mathbf{M}, \mu)$, $(\mathbf{M}, \mu) \leq (\mathbf{G}, \gamma)$. Then $(\mathbf{L}, \lambda) \leq (\mathbf{G}, \gamma)$.

The first part is a consequence of the Mackey formula for Harish-Chandra induction and restriction, while part (b) follows immediately from the transitivity of $R_{\mathbf{L}}^{\mathbf{G}}$ stated above and the fact that it sends characters to characters.

For a cuspidal pair $(\mathbf{L}, \lambda) \in \mathcal{K}$ we define its *Harish-Chandra series* to be

$$\mathrm{Irr}(\mathbf{G}^F, (\mathbf{L}, \lambda)) := \{\gamma \in \mathrm{Irr}(\mathbf{G}^F) \mid \gamma \text{ occurs in } R_{\mathbf{L}}^{\mathbf{G}}(\lambda)\}.$$

Hence Theorem 5.1(a) provides a partition

$$\mathrm{Irr}(\mathbf{G}^F) = \coprod_{(\mathbf{L}, \lambda)} \mathrm{Irr}(\mathbf{G}^F, (\mathbf{L}, \lambda))$$

of $\mathrm{Irr}(\mathbf{G}^F)$ into disjoint Harish-Chandra series where the union runs over cuspidal pairs in $\mathcal{K}$ up to $\mathbf{G}^F$-conjugation.

Much more can be said about ordinary Harish-Chandra series. For cuspidal pairs $(\mathbf{L}, \lambda) \in \mathcal{K}$ the relative Weyl group

$$W_{\mathbf{G}^F}(\mathbf{L}, \lambda) := N_{\mathbf{G}^F}(\mathbf{L}, \lambda)/\mathbf{L}^F$$

is a finite Coxeter group in a natural way. Let M be a $\mathbb{C}\mathbf{L}^F$-module affording the character λ. Howlett/Lehrer [8] have shown that the endomorphism algebra of $R_{\mathbf{L}}^{\mathbf{G}}(M)$ is an Iwahori-Hecke algebra (see Section 8 in Rouquier's lecture) of the Coxeter group $W_{\mathbf{G}^F}(\mathbf{L}, \lambda)$, possibly twisted by a 2-cocycle. Thus there is a natural bijection

$$I_{(\mathbf{L}, \lambda)}^{\mathbf{G}^F} : \mathrm{Irr}(W_{\mathbf{G}^F}(\mathbf{L}, \lambda)) \to \mathrm{Irr}(\mathbf{G}^F, (\mathbf{L}, \lambda))$$

between the set of complex irreducible characters of $W_{\mathbf{G}^F}(\mathbf{L}, \lambda)$ and the Harish-Chandra series $\mathrm{Irr}(\mathbf{G}^F, (\mathbf{L}, \lambda))$. Moreover, the decomposition of $R_{\mathbf{L}}^{\mathbf{G}}(\lambda)$ agrees with the image under $I_{(\mathbf{L}, \lambda)}^{\mathbf{G}^F}$ of the decomposition of the regular character of the relative Weyl group $W_{\mathbf{G}^F}(\mathbf{L}, \lambda)$. This statement will be made more precise in Theorem 9.1 in the general context of d-Harish-Chandra theories.

6 Generic unipotent characters

In the previous sections we have introduced the concept of generic groups to capture much of the generic nature of the subgroup structure of finite reductive groups. We now turn to the description of the generic behaviour of irreducible characters. The results in this second part are considerably deeper than those presented in the first part, hence we will not be able to say very much about the proofs.

Recall from the lecture of Geck that an irreducible character $\gamma \in \mathrm{Irr}(\mathbf{G}^F)$ is called *unipotent* if there exists an F-stable maximal torus $\mathbf{S} \leq \mathbf{G}$ such that γ is a constituent of $R_{\mathbf{S}}^{\mathbf{G}}(1)$. Let $\mathcal{U}(\mathbf{G}^F)$ denote the set of unipotent characters of $\mathbf{G}^F$. The results of Lusztig on unipotent characters sketched in Sections 9 and 10 of Geck's lecture may be rephrased as follows: there exists a set $\mathcal{U}(\mathbb{G})$ and a map

$$\mathrm{Deg} : \mathcal{U}(\mathbb{G}) \to \mathbb{Q}[x], \quad \gamma \mapsto \mathrm{Deg}(\gamma),$$

such that for any choice of p and q (and hence of $\mathbf{G}$ and F) there is a bijection $\psi_q^{\mathbb{G}} : \mathcal{U}(\mathbb{G}) \to \mathcal{U}(\mathbf{G}^F)$ such that $\psi_q^{\mathbb{G}}(\gamma)$ has degree $\psi_q^{\mathbb{G}}(\gamma)(1) = \mathrm{Deg}(\gamma)(q)$. The set $\mathcal{U}(\mathbb{G})$ is called the set of *(generic) unipotent characters of* $\mathbb{G}$.

Also recall from the lecture of Geck that for any F-stable Levi subgroup $\mathbf{L} \leq \mathbf{G}$ Deligne and Lusztig defined functors

$$R_{\mathbf{L}}^{\mathbf{G}} : \mathbb{Z}\mathrm{Irr}(\mathbf{L}^F) \to \mathbb{Z}\mathrm{Irr}(\mathbf{G}^F), \quad {}^*R_{\mathbf{L}}^{\mathbf{G}} : \mathbb{Z}\mathrm{Irr}(\mathbf{G}^F) \to \mathbb{Z}\mathrm{Irr}(\mathbf{L}^F),$$

between the character groups of $\mathbf{L}^F$ and $\mathbf{G}^F$, adjoint to each other with respect to the usual scalar product of characters. (More precisely, the definition of these functors also involves the choice of a parabolic subgroup $\mathbf{P}$ containing $\mathbf{L}$, but in our situation it turns out that they are in fact independent of this choice.) It follows from results of Shoji (see [3, Ths. 1.26 and 1.33]) that the maps $R_{\mathbf{L}}^{\mathbf{G}}$ and ${}^*R_{\mathbf{L}}^{\mathbf{G}}$ are generic:

Theorem 6.1 *For any generic Levi subgroup* $\mathbb{L}$ *of* $\mathbb{G}$ *there exist linear maps*

$$R_{\mathbb{L}}^{\mathbb{G}} : \mathbb{Z}\mathcal{U}(\mathbb{L}) \to \mathbb{Z}\mathcal{U}(\mathbb{G}), \quad {}^*R_{\mathbb{L}}^{\mathbb{G}} : \mathbb{Z}\mathcal{U}(\mathbb{G}) \to \mathbb{Z}\mathcal{U}(\mathbb{L}),$$

satisfying $\psi_q^{\mathbb{G}} \circ R_{\mathbb{L}}^{\mathbb{G}} = R_{\mathbf{L}}^{\mathbf{G}} \circ \psi_q^{\mathbb{L}}$ *for all* q *(when extending* $\psi_q^{\mathbb{G}}$ *linearly to* $\mathbb{Z}\mathcal{U}(\mathbb{G})$*).*

In this sense, the sets of unipotent characters together with the collection of functors $R_{\mathbf{L}}^{\mathbf{G}}$ and ${}^*R_{\mathbf{L}}^{\mathbf{G}}$ are generic for a series of groups of Lie type.

The Jordan decomposition of irreducible characters of $\mathbf{G}^F$ (see Section 9 in Geck's lecture) shows the prominent role of unipotent characters in the description of $\mathrm{Irr}(\mathbf{G}^F)$. We will hence restrict our attention to the subset $\mathcal{U}(\mathbf{G}^F) \subset \mathrm{Irr}(\mathbf{G}^F)$, respectively to its generic version $\mathcal{U}(\mathbb{G})$.

Example 6.2 In the case of the finite reductive group $\mathbf{G}^F = \mathrm{GL}_n(q)$ the unipotent characters are just the constituents of the permutation character $1_{\mathbf{B}^F}^{\mathbf{G}^F}$ on the F-fixed points of an F-stable Borel subgroup $\mathbf{B}$ of $\mathbf{G}$. The endomorphism algebra of this permutation module is the Iwahori–Hecke algebra of the symmetric group $\mathfrak{S}_n$ (the Coxeter group of type A_{n-1}) with parameter q (see Section 8 in the lecture of Rouquier). Its irreducible characters are in bijection with the irreducible characters of the symmetric group $\mathfrak{S}_n$, the Weyl group of $\mathbf{G}^F$. Since the latter are naturally indexed by partitions

$\alpha \vdash n$, the same is true for the unipotent characters of $\mathbf{G}^F$. Hence in this case we have $\mathcal{U}(\mathrm{GL}_n) = \{\gamma_\alpha \mid \alpha \vdash n\}$. The function $\mathrm{Deg} : \mathcal{U}(\mathrm{GL}_n) \to \mathbb{Q}[x]$ can be described as follows: for a partition $\alpha = (\alpha_1 \leq \ldots \leq \alpha_m)$ of n let $\beta_i := \alpha_i + i - 1$, $1 \leq i \leq m$. Then

$$\mathrm{Deg}(\gamma_\alpha) = \frac{(x-1)\ldots(x^n-1)\prod_{j>i}(x^{\beta_j} - x^{\beta_i})}{x^{\binom{m-1}{2}+\binom{m-2}{2}+\ldots}\prod_i \prod_{j=1}^{\beta_i}(x^j - 1)}.$$

(Despite appearance, this always is a polynomial.) Furthermore, there exists a bijection

$$\mathcal{U}(\mathrm{GL}_n) \to \mathcal{U}(\mathrm{GU}_n), \quad \gamma \mapsto \gamma^-,$$

such that $\mathrm{Deg}(\gamma^-)(x) = \pm\mathrm{Deg}(\gamma)(-x)$. This is another consequence of the Ennola duality between GL_n and GU_n.

7 d-Harish-Chandra theories

The idea that a generalized Harish-Chandra theory should exist for the unipotent characters of a finite group of Lie type first occurred in the papers by Fong/Srinivasan [7] and Schewe [11] in particular cases. The general case was settled in [3].

7.1 d-cuspidal characters

Let $\mathbb{G}$ be a generic finite reductive group and d an integer such that $a(d) > 0$, i.e., such that Φ_d divides $|\mathbb{G}|$. We first have to introduce a generalization of cuspidal characters:

Definition 7.1 A generic unipotent character $\gamma \in \mathcal{U}(\mathbb{G})$ is called *d-cuspidal* if ${}^*R_{\mathbb{L}}^{\mathbb{G}}(\gamma) = 0$ for all d-split Levi subgroups $\mathbb{L}$ properly contained in $\mathbb{G}$.

Thus in the case $d = 1$ the 1-cuspidal characters of $\mathbb{G}$ are those whose image under $\psi_q^{\mathbb{G}}$ is a cuspidal unipotent character as defined in Section 5. We have the following alternative characterization of d-cuspidality (see [3, Prop. 2.9]):

Proposition 7.2 *A unipotent character $\gamma \in \mathcal{U}(\mathbb{G})$ is d-cuspidal if and only if* $\mathrm{Deg}(\gamma)_d = |\mathbb{G}_{ss}|_d$.

Here, we write f_d for the Φ_d-part of $f \in \mathbb{Q}[x]$. The semisimple quotient $\mathbb{G}_{\mathrm{ss}}$ of $\mathbb{G}$ was defined in Section 2.

Example 7.3 We continue Example 6.2. Let $\alpha = (\alpha_1 \leq \ldots \leq \alpha_m)$ be a partition of n. A *d-hook of* α is a pair $h = (\nu, \beta)$ of integers $0 \leq \nu < \beta$ such that β occurs among the $\{\alpha_i + i - 1 \mid 1 \leq i \leq m\}$ but ν does not. The *length* of the hook h is $l(h) = \beta - \nu$ and h is also called an $l(h)$-hook. A moment's thought shows that up to a power of x the degree given in Example 6.2 of the unipotent character γ_α of the generic group of type GL_n is of the form

$$\frac{(x-1)\dots(x^n-1)}{\prod_{h \text{ hook of } \alpha}(x^{l(h)}-1)}.$$

Note that the existence of an ad-hook of α for some $a \geq 1$ implies the existence of a d-hook. Thus, by Proposition 7.2 the unipotent character γ_α is d-cuspidal if and only if α has no d-hook. Such partitions α are also called *d-cores*. In particular, GL_n has a 1-cuspidal unipotent character if and only if $n = 0$, since the empty partition is the only 1-core. This shows that indeed all unipotent characters of GL_n occur in $R_{\mathbb{S}}^{\mathbb{G}}(1)$ for the maximally split torus $\mathbb{S}$.

7.2 d-Harish-Chandra series

A pair $(\mathbb{L}, \lambda)$ consisting of a d-split Levi subgroup $\mathbb{L}$ of $\mathbb{G}$ and a unipotent character $\lambda \in \mathcal{U}(\mathbb{L})$ is called *d-split*. It is called *d-cuspidal* if moreover λ is d-cuspidal. We introduce the following relation on the set of d-split pairs:

Definition 7.4 Let $(\mathbb{M}_1, \mu_1)$ and $(\mathbb{M}_2, \mu_2)$ be d-split in $\mathbb{G}$. Then we say that $(\mathbb{M}_1, \mu_1) \leq_d (\mathbb{M}_2, \mu_2)$ if $\mathbb{M}_1$ is a d-split Levi subgroup of $\mathbb{M}_2$ and μ_2 occurs in $R_{\mathbb{M}_1}^{\mathbb{M}_2}(\mu_1)$.

For a d-cuspidal pair $(\mathbb{L}, \lambda)$ of $\mathbb{G}$ we write

$$\mathcal{U}(\mathbb{G}, (\mathbb{L}, \lambda)) := \{\gamma \in \mathcal{U}(\mathbb{G}) \mid (\mathbb{L}, \lambda) \leq_d (\mathbb{G}, \gamma)\}$$

for the set of unipotent characters of $\mathbb{G}$ lying above $(\mathbb{L}, \lambda)$. We call $\mathcal{U}(\mathbb{G}, (\mathbb{L}, \lambda))$ the *d-Harish-Chandra series above* $(\mathbb{L}, \lambda)$ because of the following fundamental result, which shows that for any d we obtain a generalized Harish-Chandra theory (in generalization of the case $d = 1$ in Theorem 5.1):

Theorem 7.5 *Let $\mathbb{G}$ be a generic finite reductive group and $d \geq 1$ such that Φ_d divides $|\mathbb{G}|$.*

(a) (Disjointness) The sets $\mathcal{U}(\mathbb{G}, (\mathbb{L}, \lambda))$ (where $(\mathbb{L}, \lambda)$ runs over a system of representatives of the $W_{\mathbb{G}}$-conjugacy classes of d-cuspidal pairs) form a partition of $\mathcal{U}(\mathbb{G})$.

(b) (Transitivity) Let $(\mathbb{L}, \lambda)$ be d-cuspidal and $(\mathbb{M}, \mu)$ be d-split such that $(\mathbb{L}, \lambda) \leq_d (\mathbb{M}, \mu)$ and $(\mathbb{M}, \mu) \leq_d (\mathbb{G}, \gamma)$. Then $(\mathbb{L}, \lambda) \leq_d (\mathbb{G}, \gamma)$.

The statement of Theorem 7.5 has been checked case by case (except for the case $d = 1$ where a conceptual proof is known). It is a consequence of the more precise Theorem 9.1 about the decomposition of twisted induction of unipotent characters from d-split Levi subgroups. Note that, in contrast to the situation for Theorem 5.1, part (b) is far from obvious since $R_{\mathbb{L}}^{\mathbb{G}}$ does not necessarily take characters to actual characters.

Example 7.6 We continue Example 7.3, where we saw that the d-cuspidal characters of GL_n are indexed by partitions of n which are d-cores. Let $h = (\nu, \beta)$ be a d-hook of the partition $\alpha = (\alpha_1 \leq \ldots \leq \alpha_m)$, and let j be the index with $\alpha_j + j - 1 = \beta$. Then $\alpha' = (\alpha'_1 \leq \ldots \leq \alpha'_m)$ is called the partition obtained from α by *removing the d-hook h* if the set $\{\alpha'_i + i - 1\}$ coincides with the set

$$\{\alpha_i + i - 1 \mid i \neq j\} \cup \{\alpha_j + j - 1 - (\beta - \nu)\}.$$

The d-core obtained from α by successively removing all possible d-hooks is called the *d-core of* α. (The reader may want to check that the d-core does not depend on the order in which d-hooks are removed and is hence well-defined.) It can be shown that the d-Harish-Chandra series of GL_n above the d-cuspidal character indexed by the d-core α consists of the unipotent characters indexed by partitions of n whose d-core is α.

In Example 3.3(b) we determined the structure of d-split Levi subgroups of GL_n. By Example 7.3 the generic group $\mathrm{GL}_{n_i}^{(d)}$ has a d-cuspidal unipotent character only if $n_i = 1$. Thus the d-cuspidal pairs of GL_n are precisely the pairs $(\mathbb{L}, \lambda_\alpha)$ where

$$\mathbb{L} = \mathrm{GL}_1^{(d)} \times \ldots \times \mathrm{GL}_1^{(d)} \times \mathrm{GL}_s$$

(r factors $\mathrm{GL}_1^{(d)}$) with $n = dr + s$, and λ_α is the unipotent character of $\mathbb{L}$ parametrized by the d-core $\alpha \vdash s$.

8 Generic blocks

One importance of d-Harish-Chandra series lies in their connection with ℓ-blocks of finite groups of Lie type for primes ℓ not dividing q. Before continuing the exposition of the generic theory we therefore briefly explain this application.

We fix $\mathbb{G}$ and a choice of a prime power q, hence a pair $(\mathbf{G}, F)$. A prime ℓ not dividing q is called *large for* $\mathbb{G}$ if ℓ does not divide the order of $W\langle\phi\rangle$. If ℓ is large then by Proposition 4.3(a) there exists a unique d such that $\ell | \Phi_d(q)$ and $\Phi_d \big\| |\mathbb{G}|$.

Theorem 8.1 (Broué/Malle/Michel) *Let ℓ be large for $\mathbb{G}$, and assume that $\ell | \Phi_d(q)$, $\Phi_d \big\| |\mathbb{G}|$. Then the partition of $\mathcal{U}(\mathbf{G}^F)$ into ℓ-blocks coincides with the image under $\psi_q^{\mathbb{G}}$ of the partition of $\mathcal{U}(\mathbb{G})$ into d-Harish-Chandra series. In particular, the distribution of unipotent characters of $\mathbf{G}^F$ into ℓ-blocks is generic.*

This is proved in [3, Th. 5.24]. Let us write $b_\ell(\mathbf{L},\lambda)$ for the unipotent ℓ-block of $\mathbf{G}^F$ indexed by $(\mathbb{L},\lambda)$. The unipotent blocks and their defect groups can be described more precisely. For this we need:

Definition 8.2 For $\gamma \in \mathcal{U}(\mathbb{G})$ let $S_d(\gamma)$ denote the set of d-tori of $\mathbb{G}$ contained in a maximal torus $\mathbb{S}$ of $\mathbb{G}$ such that ${}^*R_{\mathbb{S}}^{\mathbb{G}}(\gamma) \neq 0$. The maximal elements of $S_d(\gamma)$ are called the *d-defect tori of* γ.

The d-defect tori of a unipotent character can be characterized as follows (see [3, Th. 4.8]):

Proposition 8.3 *Let* $(\mathbb{L},\lambda)$ *be a d-cuspidal pair and* $\gamma \in \mathcal{U}(\mathbb{G},(\mathbb{L},\lambda))$. *Then the d-defect tori of* γ *are conjugate to* $\mathrm{Rad}(\mathbb{L})_d$.

Here $\mathrm{Rad}(\mathbb{L})_d$ denotes the Sylow d-torus of the torus $\mathrm{Rad}(\mathbb{L})$. For an F-stable Levi subgroup $\mathbf{L}$ of $\mathbf{G}$ denote by $\mathrm{Ab}_\ell\mathrm{Irr}(\mathbf{L}^F)$ the group of characters (over a splitting field of characteristic 0) of ℓ-power order of the abelian group $\mathbf{L}^F/[\mathbf{L}^F,\mathbf{L}^F]$. Then [3, Th. 5.24] gives the following sharpening of Theorem 8.1:

Theorem 8.4 (Broué/Malle/Michel) *Let* ℓ *be large for* $\mathbb{G}$, *and assume that* $\ell|\Phi_d(q)$, $\Phi_d\big||\mathbb{G}|$. *Let* $(\mathbb{L},\lambda)$ *be a d-cuspidal pair of* $\mathbb{G}$.

(a) The ℓ*-block* $b_\ell(\mathbf{L},\lambda)$ *of* $\mathbf{G}^F$ *consists of the irreducible constituents of the virtual characters* $R_{\mathbf{L}}^{\mathbf{G}}(\theta\lambda)$, *where* $\theta \in \mathrm{Ab}_\ell\mathrm{Irr}(\mathbf{L}^F)$.

(b) The defect groups of $b_\ell(\mathbf{L},\lambda)$ *are the Sylow* ℓ*-subgroups of the groups of F-fixed points of the d-defect tori of unipotent characters in* $b_\ell(\mathbf{L},\lambda)$.

The structure of the Sylow ℓ-subgroup of the group of F-fixed points of a d-torus was described in Proposition 4.3(c).

Example 8.5 Theorem 8.4 and the description of d-Harish-Chandra series for GL_n in Example 7.6 provide the following description of unipotent ℓ-blocks of $\mathrm{GL}_n(q)$ for large primes ℓ dividing $\Phi_d(q)$. Two unipotent characters γ_α, $\gamma_{\alpha'}$ lie in the same ℓ-block if and only if the d-cores of α and α' coincide. (This was first proved by Fong and Srinivasan.) This result is very much reminiscent of the description of ℓ-blocks in the symmetric group $\mathfrak{S}_n$.

9 Relative Weyl groups

The d-Harish-Chandra theories presented in Section 7 seem to be just the shadow of a much deeper theory describing the decomposition of the functor of twisted induction. Much of this is still conjectural.

9.1 The decomposition of $R_{\mathbb{L}}^{\mathbb{G}}$

The only known proof of Theorem 7.5 in the case $d \neq 1$ consists in determining the decomposition of the Deligne-Lusztig induced characters of d-cuspidal unipotent characters. To state this result we need to introduce an important invariant of a d-Harish-Chandra series. Let $(\mathbb{L}, \lambda)$ be a d-cuspidal pair in $\mathbb{G}$. Let $(\mathbf{G}, F)$ be a finite reductive group associated to $\mathbb{G}$, and let $\mathbf{L}$ be an F-stable Levi subgroup of $\mathbf{G}$ corresponding to $\mathbb{L}$ according to Proposition 2.2(b). Then the relative Weyl group $W_{\mathbf{G}^F}(\mathbf{L}) = N_{\mathbf{G}^F}(\mathbf{L})/\mathbf{L}^F$ acts on $\mathrm{Irr}(\mathbf{L}^F)$. By results of Lusztig, this leaves the subset $\mathcal{U}(\mathbf{L}^F)$ of unipotent characters invariant. Moreover, the action on $\mathcal{U}(\mathbf{L}^F)$ is generic in the sense that it is possible to define an action of $W_{\mathbb{G}}(\mathbb{L}) = N_{W_{\mathbb{G}}}(W_{\mathbb{L}})/W_{\mathbb{L}}$ on $\mathcal{U}(\mathbb{L})$ which, under all $\psi_q^{\mathbb{G}}$, specializes to the action of $W_{\mathbf{G}^F}(\mathbf{L})$ on $\mathcal{U}(\mathbf{L}^F)$ (see [3]). This gives sense to the definition

$$W_{\mathbb{G}}(\mathbb{L}, \lambda) := N_{W_{\mathbb{G}}}(W_{\mathbb{L}}, \lambda)/W_{\mathbb{L}}$$

of the *relative Weyl group of* $(\mathbb{L}, \lambda)$ *in* $\mathbb{G}$.

Theorem 9.1 (Broué/Malle/Michel) *For each d there exists a collection of isometries*

$$I_{(\mathbb{L},\lambda)}^{\mathbb{M}} : \mathbb{Z}\mathrm{Irr}(W_{\mathbb{M}}(\mathbb{L}, \lambda)) \to \mathbb{Z}\mathcal{U}(\mathbb{M}, (\mathbb{L}, \lambda)),$$

such that for all $\mathbb{M}$ *and all* $(\mathbb{L}, \lambda)$ *we have*

$$R_{\mathbb{M}}^{\mathbb{G}} \circ I_{(\mathbb{L},\lambda)}^{\mathbb{M}} = I_{(\mathbb{L},\lambda)}^{\mathbb{G}} \circ \mathrm{Ind}_{W_{\mathbb{M}}(\mathbb{L},\lambda)}^{W_{\mathbb{G}}(\mathbb{L},\lambda)}.$$

Here $\mathbb{M}$ *runs over the d-split Levi subgroups of* $\mathbb{G}$ *and* $(\mathbb{L}, \lambda)$ *over the set of d-cuspidal pairs of* $\mathbb{M}$.

This was proved in [3, Th. 3.2] by using results of Asai on the decomposition of $R_{\mathbf{L}}^{\mathbf{G}}$ in the case of classical groups, and by explicit determination of these decompositions in the case of exceptional groups.

An isometry I from $\mathbb{Z}\mathrm{Irr}(W_{\mathbb{G}}(\mathbb{L}, \lambda))$ to $\mathbb{Z}\mathcal{U}(\mathbb{G}, (\mathbb{L}, \lambda))$ is nothing else but a bijection

$$\mathrm{Irr}(W_{\mathbb{G}}(\mathbb{L}, \lambda)) \xrightarrow{\sim} \mathcal{U}(\mathbb{G}, (\mathbb{L}, \lambda)), \quad \chi \mapsto \gamma_\chi,$$

together with a collection of signs

$$\{\epsilon(\gamma) \mid \gamma \in \mathcal{U}(\mathbb{G}, (\mathbb{L}, \lambda))\},$$

such that

$$I(\chi) = \epsilon(\gamma_\chi)\, \gamma_\chi\,.$$

Thus, Theorem 9.1 states that up to an adjustment by suitable signs, twisted induction from d-split Levi subgroups is nothing but ordinary induction in relative Weyl groups. As a consequence of [3, Th. 5.24] we moreover have the following congruence of character degrees

$$\epsilon(\gamma_\chi)\,\mathrm{Deg}(\gamma_\chi) \equiv \chi(1) \pmod{\Phi_d} \quad \text{in } \mathbb{Q}[x]$$

in the situation of Theorem 9.1.

Example 9.2 We continue the Example 4.2 of generic groups of type ${}^3D_4$. The relative Weyl groups $W_{\mathbb{G}}(\mathbb{S}_d)$ for the Sylow d-tori $\mathbb{S}_d$ with $d \in \{3,6\}$ turn out to be isomorphic to $\mathrm{SL}_2(3)$. Since $\mathrm{SL}_2(3)$ has 7 irreducible characters Theorem 9.1 implies that $\mathcal{U}(\mathbb{G},(\mathbb{S}_d,1))$ has cardinality 7 for $d \in \{3,6\}$. Moreover, the vector of degrees of the irreducible characters of $\mathrm{SL}_2(3)$ is $(1,1,1,2,2,2,3)$, so $R_{\mathbb{S}_d}^{\mathbb{G}}(1)$ contains three constituents with multiplicity ± 1, three with multiplicity ± 2 and one with multiplicity ± 3.

9.2 Relative Weyl groups are pseudo-reflection groups

Theorem 9.1 turns attention towards the relative Weyl groups of d-cuspidal pairs $(\mathbb{L},\lambda)$. We stated in Section 5 that in the case $d = 1$ the relative Weyl groups turn out to be Coxeter groups, that is, groups with a faithful representation as a group generated by real reflections. In a case-by-case analysis the following surprising fact can be verified (see [2]):

Proposition 9.3 *Let $\mathbb{G}$ be a generic group and d a positive integer. For any d-cuspidal pair $(\mathbb{L},\lambda)$ of $\mathbb{G}$ the relative Weyl group $W_{\mathbb{G}}(\mathbb{L},\lambda)$ is a pseudo-reflection group, i.e., a finite group having a faithful representation as a group generated by complex reflections.*

(See Section 9 of Rouquier's talk for an introduction to pseudo-reflection groups.) In the case that $\mathbb{L}$ is a maximal torus (and hence $\lambda = 1$) this is a result of Springer [12], but no a-priori proof is known in the general case.

Remark 9.4 It even turns out that, if W acts irreducibly on $V = Y \otimes_{\mathbb{Z}} \mathbb{C}$, then the pseudo-reflection groups $W_{\mathbb{G}}(\mathbb{L},\lambda)$ are also irreducible in their natural reflection representation.

Example 9.5 Let $\mathbb{G}$ be a generic group of type E_7 and let $d = 4$. A Sylow 4-torus $\mathbb{S}$ of $\mathbb{G}$ has order Φ_4^2 and its centralizer is a 4-split Levi subgroup $\mathbb{L} = C_{\mathbb{G}}(\mathbb{S})$ with semisimple part $(\mathrm{PGL}_2)^3$. Since $\mathbb{L}$ is minimal 4-split all its unipotent characters are 4-cuspidal. The relative Weyl group $W_{\mathbb{G}}(\mathbb{L})$ is a two-dimensional complex reflection group of order 96, denoted G_8 by Shephard and Todd. It has two orbits of length 3 and two fixed points on the set $\mathcal{U}(\mathbb{L})$. In particular, the relative Weyl group $W_{\mathbb{G}}(\mathbb{L},\lambda)$ for a (4-cuspidal) character λ in one of the orbits of length 3 is strictly smaller than $W_{\mathbb{G}}(\mathbb{L})$. It turns out to be the imprimitive complex reflection group denoted $G(4,1,2)$ of order 32.

The relative Weyl groups occurring in generic finite reductive groups of exceptional type are collected in [3, Tables 1 and 3] while those for classical types are described in [3, Sec. 3], see also [2, 3B].

9.3 Cyclotomic Hecke algebras

We saw in Section 5 that in the ordinary Harish-Chandra case $d = 1$, the assertion of Theorem 9.1 has the following explanation. The endomorphism algebra of a Harish-Chandra induced cuspidal module with character $\lambda \in \mathrm{Irr}(\mathbf{L}^F)$ is a Hecke algebra, a deformation of the group algebra of the relative Weyl group $W_{\mathbf{G}^F}(\mathbf{L}, \lambda)$. This yields an indexation of the Harish-Chandra series $\mathrm{Irr}(\mathbf{G}^F, (\mathbf{L}, \lambda))$ by $\mathrm{Irr}(W_{\mathbf{G}^F}(\mathbf{L}, \lambda))$. It is tempting to conjecture that for arbitrary d a similar deep reason lies behind the result of Theorem 9.1. A conjecture in this direction was put forward in [2]. The only situation where this general conjecture is known to be true is that where $\mathbb{L}$ is a Coxeter torus of $\mathbb{G}$, by work of Lusztig [9]. Apart from that case, at present only some consequences of this deep conjecture have been verified.

More precisely, in the paper [2] we introduced certain deformations of group algebras of complex reflection groups, called cyclotomic Hecke algebras. Given an irreducible complex reflection group W, there are two possible definitions for the cyclotomic algebra $\mathcal{H}(W)$. The first is by generators and relations obtained by deforming a Coxeter like presentation of W. This approach, which is analogous to the usual definition of the Iwahori-Hecke algebra attached to a finite Coxeter group, was taken in [2]. The alternative is the construction of $\mathcal{H}(W)$ as a natural quotient of the group algebra of the topological braid group $\mathcal{B}_W$ associated to W, i.e., of the fundamental group of the space of regular orbits of W in its reflection representation (see Section 10 in Rouquier's lecture). The equivalence of the two definitions was shown by Deligne and Brieskorn in the case that G is a real reflection group, and by Broué/Malle/Rouquier [4] in almost all complex cases.

Conjecturally, these cyclotomic algebras should be the correct replacement for Iwahori-Hecke algebras in the explanation of the result of Theorem 9.1 for the case $d = 1$. They form the object of some current research, see for example [4] and [10] and the references cited there. Unfortunately, we cannot go into more details about this fascinating subject.

References

[1] M. Broué and G. Malle, Théorèmes de Sylow génériques pour les groupes réductifs sur les corps finis, *Math. Ann.* **292** (1992), 241–262.

[2] M. Broué and G. Malle, Zyklotomische Heckealgebren, in: *Représentations unipotentes génériques et blocs des groupes réductifs finis, Astérisque* **212** (1993), 119–189.

[3] M. Broué, G. Malle and J. Michel, Generic blocks of finite reductive groups, in: *Représentations unipotentes génériques et blocs des groupes réductifs finis, Astérisque* **212** (1993), 7–92.

[4] M. Broué, G. Malle and R. Rouquier, Complex reflection groups, braid groups, Hecke algebras, to appear in *J. reine angew. Math.*

[5] R. W. Carter, *Simple groups of Lie type*, Wiley, London, 1972.

[6] F. Digne and J. Michel, *Representations of finite groups of Lie type*, London Math. Soc. Students Texts **21**, Cambridge University Press, Cambridge, 1991.

[7] P. Fong and B. Srinivasan, Generalized Harish–Chandra theory for unipotent characters of finite classical groups, *J. Algebra* **104** (1986), 301–309.

[8] R. B. Howlett and G. I. Lehrer, Induced cuspidal representations and generalized Hecke rings, *Invent. Math.* **58** (1980), 37–64.

[9] G. Lusztig, Coxeter orbits and eigenspaces of Frobenius, *Invent. Math.* **38** (1976), 101–159.

[10] G. Malle and A. Mathas, Symmetric cyclotomic Hecke algebras, to appear in *J. Algebra.*

[11] K.-D. Schewe, *Blöcke exzeptioneller Chevalley-Gruppen*, Dissertation, Bonner Mathematische Schriften, nr. 165, Bonn, 1985.

[12] T. A. Springer, Regular elements of finite reflection groups, *Invent. Math.* **25** (1974), 159–198.

Introduction to Quantum Groups

Jens Carsten Jantzen

Matematisk Institut, Aarhus Universitet, DK-8000 Aarhus C, Denmark

Quantum groups were introduced independently by Drinfel'd in [Dr] and by Jimbo in [Ji] when dealing with problems in mathematical physics. By now these objects have found applications in many areas of mathematics. Two of them are most important within the context of this conference. One of them is the discovery of certain bases of irreducible representations of complex semisimple Lie algebras (and algebraic groups) with certain remarkable properties; see 3.6 below. The other one is the discovery of algebras in characteristic 0 whose representation theory is very similar to that of algebraic groups in prime characteristic. This phenomenon is one of the key ingredients in Lusztig's programme to attack his conjectured character formula for algebraic groups in prime characteristic (described in Donkin's lectures in this volume).

The topics in my lectures were chosen with the aim to provide some background for these applications. For the proofs I usually refer to my book [LQG] where the choice of topics was guided by the same principles. More information can be found in the books [CP], [Jo], [Ka], [Lu2] that cover additional topics and work in greater generality.

Lecture 1: Quantum $\mathfrak{sl}_2$

The goal of this first lecture is to discuss an example that is typical for the general construction later on. Fix throughout this lecture an arbitrary ground field k. All associative algebras are supposed to have an identity.

1.1. The Lie algebra $\mathfrak{sl}_2$ of all (2×2)–matrices over k with trace 0 has basis e, f, h where

$$e = \begin{pmatrix} 0 & 1 \\ 0 & 0 \end{pmatrix}, \qquad f = \begin{pmatrix} 0 & 0 \\ 1 & 0 \end{pmatrix}, \qquad h = \begin{pmatrix} 1 & 0 \\ 0 & -1 \end{pmatrix}. \tag{1}$$

The universal enveloping algebra $U(\mathfrak{sl}_2)$ of $\mathfrak{sl}_2$ can then be described as the associative algebra over k with generators e, f, and h and relations

$$he - eh = 2e, \qquad hf - fh = -2f, \qquad ef - fe = h. \tag{2}$$

We are now going to define a "quantum deformation" U_q of this enveloping algebra. It depends on a parameter $q \in k$, which can be any element in k

satisfying $q \neq 0$ and $q^2 \neq 1$. Then U_q is defined as the associative algebra over k with generators E, F, K, and K^{-1} and relations

$$KK^{-1} = 1 = K^{-1}K, \tag{3}$$
$$KEK^{-1} = q^2 E, \tag{4}$$
$$KFK^{-1} = q^{-2} F, \tag{5}$$
$$EF - FE = \frac{K - K^{-1}}{q - q^{-1}}. \tag{6}$$

We call this algebra the *quantised enveloping algebra* of $\mathfrak{sl}_2$ or just "quantum $\mathfrak{sl}_2$".

The general feeling that you should have about the relation between U_q and $U(\mathfrak{sl}_2)$ is that E and F correspond to e and f while K is something like an exponential in h involving the parameter q.

There are other definitions of U_q that lead to similar algebras that share many properties with the version of U_q studied here.

1.2. In the next subsections I want to point out that U_q has many features that are similar to those of $U(\mathfrak{sl}_2)$, but that there are also significant differences.

Let me start with looking at bases. The Poincaré-Birkhoff-Witt (PBW) theorem (cf. [Ca], Section 7 or [Hu], 17.3) says that all $f^r h^m e^s$ with m, r, s non-negative integers are a basis of $U(\mathfrak{sl}_2)$. We have similarly:

Theorem: *The monomials $F^s K^n E^r$ with $r, s, n \in \mathbf{Z}$ and $r, s \geq 0$ are a basis of U_q.*

It is rather straightforward to show (using the defining relations from 1.1) that these elements span U_q. One can prove their linear independence by constructing an explicit action of U_q on some vector space, cf. [LQG], 1.5.

The enveloping algebra of any Lie algebra has a natural filtration. The PBW theorem implies then that the associated graded algebra is isomorphic to the symmetric algebra over the Lie algebra, cf. [Hu], 17.3. This in turn implies that the enveloping algebra has no zero divisors. With a little modification this argument works also in our situation and yields, cf. [LQG], 1.8.

Proposition: *The algebra U_q has no zero divisors.*

1.3. The next aspect of these rings that I want to compare are the centres. If k has characteristic 0, then the centre of $U(\mathfrak{sl}_2)$ is generated (as an algebra over k) by the "Casimir element" $c = 4fe + h^2 + 2h$. If k has prime characteristic p, then also e^p, f^p, and $h^p - h$ are in the centre of $U(\mathfrak{sl}_2)$; if $p > 2$ then these three elements together with c generate the centre. (Cf. [RŠ], Conjecture 1; note that *conjecture* is a rather unconventional translation of предложение.)

Let us now look at U_q. We have here an analogue to the Casimir element, namely

$$C = FE + \frac{Kq + K^{-1}q^{-1}}{(q - q^{-1})^2}. \tag{1}$$

It is always contained in the centre of U_q. We get now, cf. [LQG], 2.18, 2.20:

Proposition: a) *If q is not a root of unity, then the centre of U_q is generated by C.*

b) *If q is a primitive l–th root of unity with l odd, $l \geq 3$, then the centre of U_q is generated by E^l, F^l, K^l, K^{-l}, and C.*

We see that here the behaviour of U_q does not depend so much on the characteristic p, but on q. If q is not a root of unity, then U_q behaves (for any k) as $U(\mathfrak{sl}_2)$ does in characteristic 0, if q is a root of unity, then U_q behaves as $U(\mathfrak{sl}_2)$ in prime characteristic. We shall encounter the same phenomenon when we look in the next subsections at finite dimensional representations.

1.4. All finite dimensional representations of the Lie algebra $\mathfrak{sl}_2$ (hence: all finite dimensional $U(\mathfrak{sl}_2)$–modules) are well known, cf. [Hu], 7.2. To start with, they are completely reducible; so it suffices to know the irreducible representations. Furthermore there exists for each integer $n \geq 0$ exactly one (up to isomorphism) simple module $L(n)$ of dimension $n+1$; this module has a basis $(v_i \mid 0 \leq i \leq n)$ such that $hv_i = (n-2i)v_i$ and

$$v_i = \begin{cases} v_{i+1}, & \text{if } i < n, \\ 0, & \text{if } i = n, \end{cases} \qquad ev_i = \begin{cases} i(n+1-i)v_{i-1}, & \text{if } i > 0, \\ 0, & \text{if } i = 0. \end{cases}$$

The quantum version of the first part of this result is (cf. [LQG], 2.9, 2.3):

Proposition: *Suppose that q is not a root of unity. If the characteristic of k is different from 2, then every finite dimensional U_q–module is semisimple.*

The exclusion of the case where k has characteristic 2 is necessary: We can in that case construct modules where E and F act as 0, and where K acts via a unipotent linear map of order 2. However, if we consider only U_q–modules that are the direct sum of their weight spaces, then the conclusion of the proposition holds also in characteristic 2. Here the *weight spaces* of a U_q–module V are defined as the

$$V_\lambda = \{ v \in V \mid Kv = \lambda v \}, \tag{1}$$

with $\lambda \in k$.

We turn next to an explicit description of the simple U_q–modules. This will involve certain elements that we denote by $[n]$; we set for all integers n

$$[n] = \frac{q^n - q^{-n}}{q - q^{-1}}. \tag{2}$$

For example, we have $[1] = 1$ and $[2] = q + q^{-1}$. In general, $[n]$ is a polynomial in q and q^{-1} with integer coefficients such that one gets n, when one replaces q by 1. Using this notation, we get (cf. [LQG], 2.6):

Theorem: *Suppose that q is not a root of unity. There are for each integer $n \geq 0$ a simple U_q–module $L(n, +)$ with basis $v_0, v_1, \ldots, v_n$ such that $Kv_i = q^{n-2i} v_i$ and*

$$Fv_i = \begin{cases} v_{i+1}, & \text{if } i < n, \\ 0, & \text{if } i = n, \end{cases} \qquad Ev_i = \begin{cases} [i][n+1-i]v_{i-1}, & \text{if } i > 0, \\ 0, & \text{if } i = 0, \end{cases}$$

and a simple U_q–module $L(n, -)$ with basis $v'_0, v'_1, \ldots, v'_n$ such that $Kv'_i = -q^{n-2i} v'_i$ and

$$Fv'_i = \begin{cases} v'_{i+1}, & \text{if } i < n, \\ 0, & \text{if } i = n, \end{cases} \qquad Ev'_i = \begin{cases} -[i][n+1-i]v'_{i-1}, & \text{if } i > 0, \\ 0, & \text{if } i = 0. \end{cases}$$

Each simple U_q–module of dimension $n+1$ is isomorphic to $L(n, +)$ or to $L(n, -)$.

Let me say something about the proof of the last claim. Assume for the sake of simplicity that k is algebraically closed. Let V be a simple U_q–module. The element K acting on V has at least one eigenvalue (say $\lambda \in k$), since k is algebraically closed. Using 1.1(4),(5) one checks easily that $EV_\mu \subset V_{\mu q^2}$ and $FV_\mu \subset V_{\mu q^{-2}}$. This implies that the (direct) sum of all $V_{\lambda q^{2i}}$ with $i \in \mathbf{Z}$ is a non-zero submodule of V, hence all of V.

Since V is finite dimensional, there exists i maximal with $V_{\lambda q^{2i}} \neq 0$. After renaming λ we can assume that $V_\lambda \neq 0$ while $V_{\lambda q^2} = 0$. Choose $v_0 \in V_\lambda$, $v_0 \neq 0$. Set $v_i = F^i v_0$ for all $i > 0$. We have then $v_i \in V_{\lambda q^{-2i}}$ for all i, i.e., $Kv_i = \lambda q^{-2i} v_i$. In particular the non-zero v_i are linearly independent. So there are only finitely many i with $v_i \neq 0$, and we can find j with $v_j \neq 0$ and $v_{j+1} = 0$.

We have $Ev_0 \in V_{\lambda q^2} = 0$, hence $Ev_0 = 0$. Using 1.1(6) one can now compute for each $i > 0$ an element $c_i(\lambda) \in k$ with $Ev_i = c_i(\lambda) v_i$. I leave that computation as an exercise. Now $v_{j+1} = 0 \neq v_j$ implies $c_{j+1}(\lambda) = 0$. Using the explicit form of $c_{j+1}(\lambda)$, one gets then $\lambda^2 = q^{2j}$, hence $\lambda = q^j$ or $\lambda = -q^j$. The first possibility leads to $L(n, +)$, the second one to $L(n, -)$.

In characteristic 2 the simple modules $L(n, +)$ and $L(n, -)$ are obviously isomorphic. In all other characteristics K has different eigenvalues on these modules. Therefore they are not isomorphic. So for characteristic $\neq 2$ the representation theory of U_q (with q not a root of unity) is very similar to that of $U(\mathfrak{sl}_2)$ in characteristic 0, with one significant change: Instead of one simple module in each dimension, there are now two.

1.5. Let me now compare simple modules for $U(\mathfrak{sl}_2)$ in case k has prime characteristic p to simple U_q–modules in case q is a root of unity. For the sake of simplification, let me assume that k is algebraically closed (in both cases) and that q is a primitive l–th root of unity with l odd and $l \geq 3$.

Recall that in the first case e^p, f^p, and $h^p - h$ are contained in the centre of $U(\mathfrak{sl}_2)$, while in the second case E^l, F^l, and K^l are in the centre of U_q. Schur's lemma tells us that these elements have to act as scalars on the simple modules. It makes sense to collect the simple modules where these scalars have the same value.

The most natural case (in some sense) for $U(\mathfrak{sl}_2)$ is that where e^p, f^p, and $h^p - h$ all act as 0. The representations of $\mathfrak{sl}_2$ with this property are usually called the "restricted" representations. They include all representations that one gets by differentiating representations of the algebraic group SL_2. It turns out that there are p simple modules (up to isomorphism) of this type, one of dimension $n+1$ for each integer n with $0 \leq n < p$. More explicitly, the module of dimension $n+1$ has a basis $(v_i \mid 0 \leq i \leq n)$ such that the action of e, f, and h on this basis is given by the formulas preceding the proposition in 1.4.

The corresponding case for U_q is that E^l and F^l act as 0 while K^l acts as 1. We can use the formulas from the theorem in 1.4 to define for all integers $n \geq 0$ a U_q–modules $L(n, +)$. If $n < l$, then these modules are simple and they are annihilated by E^l, F^l, and $K^l - 1$. We get thus all simple U_q–modules with this property. Similarly, the simple U_q–modules on which E^l and F^l act as 0 while K^l acts as -1, can be realised as the $L(n, -)$ with $0 \leq n < l$.

Consider next the case where e^p and f^p act as 0 while $h^p - h$ acts as some $a \neq 0$, $a \in k$. The corresponding quantum case is that E^l and F^l act as 0 while K^l acts as some $\alpha \in k$ with $\alpha \neq 0$ and $\alpha^2 \neq 1$. In the Lie algebra case we get for each $b \in k$ with $b^p - b = a$ a simple module of dimension p with basis $(v_i \mid 0 \leq i < p)$ such that the action is given by $hv_i = (b - 2i)v_i$ and

$$ev_i = \begin{cases} 0, & \text{if } i = 0, \\ i(b+1-i)v_{i-1}, & \text{if } i > 0, \end{cases} \qquad fv_i = \begin{cases} v_{i+1}, & \text{if } i < p, \\ 0, & \text{if } i = p. \end{cases}$$

We get in the quantum case for each $\beta \in k$ with $\beta^l = \alpha$ a simple module of dimension l with basis $(v_i \mid 0 \leq i < l)$ such that the action is given by $Kv_i = \beta q^{-2i} v_i$ and

$$Ev_i = \begin{cases} 0, & \text{if } i = 0, \\ [i]\,[\beta; 1-i]\,v_{i-1}, & \text{if } i > 0, \end{cases} \qquad Fv_i = \begin{cases} v_{i+1}, & \text{if } i < l, \\ 0, & \text{if } i = l, \end{cases}$$

where

$$[\beta; m] = \frac{\beta q^m - \beta^{-1} q^{-m}}{q - q^{-1}}. \tag{1}$$

Distinct choices of b (or β) lead to non-isomorphic modules. The number of possible b is equal to p; if b is one possibility, then the $b + i$ with $0 < i < p$

are the others. Similarly. the number of possible β is equal to l; if β is one possibility, then the βq^i with $0 < i < l$ are the others. We get thus p (or l) isomorphism classes of simple modules; they are all classes in this case.

Let me look at one additional case in detail. Suppose that e^p and $h^p - h$ act as 0 while f^p acts as 1. In the quantum case suppose that E^l acts as 0 while K^l and F^l act as 1. We get now $(p+1)/2$ resp. $(l+1)/2$ isomorphism classes of such modules. More explicitly, we get for each integer n with $0 \le n < p$ a simple module for $U(\mathfrak{sl}_2)$ with basis $(v_i \mid 0 \le i < p)$ and action given by $hv_i = (n-2i)v_i$ and

$$ev_i = \begin{cases} 0, & \text{if } i = 0, \\ i(n+1-i)v_{i-1}, & \text{if } i > 0, \end{cases} \qquad fv_i = \begin{cases} v_{i+1}, & \text{if } i < p, \\ v_0, & \text{if } i = p. \end{cases}$$

In the quantum case we get for each of these n a simple module with basis again denoted by $(v_i \mid 0 \le i < l)$ such that the action is given by $Kv_i = q^{n-2i}v_i$ and

$$Ev_i = \begin{cases} 0, & \text{if } i = 0, \\ [i]\,[n+1-i]\,v_{i-1}, & \text{if } i > 0, \end{cases} \qquad Fv_i = \begin{cases} v_{i+1}, & \text{if } i < l, \\ v_0, & \text{if } i = l, \end{cases}$$

One checks that the module for n is isomorphic to the module for $p-2-n$ (resp. for $l-2-n$) if $n \le p-2$ (resp. if $n \le l-2$) and that there are not any other isomorphisms (*Exercise!*).

The three cases discussed above are quite typical for the strong similarities between the representation theory of Lie algebras in prime characteristic and that of quantum groups "at a root of unity" — and not just in the case of $\mathfrak{sl}_2$. There will not be enough time to discuss this in the general case. Let me refer you to papers by De Concini and Kac (and Procesi) discussing the quantum case and exhibiting strong analogies with the older Lie algebra case: see [DK1], [DK2], [DKP]. Let me also mention that these quantum groups play a crucial role in Lusztig's programme for attacking the Lusztig conjecture, cf. Donkin's lectures [Do] in these proceedings.

1.6. In the last two subsections, I have emphasised similarities between the representations of the Lie algebra $\mathfrak{sl}_2$ and the theory of U_q–modules. However, there are crucial constructions (such as tensor products and contragedient representations) in the representation theory of any Lie algebra, for which we do not yet have seen a U_q–analogue. That will change now.

The tensor product $M \otimes N$ of two U_q–modules M and N is to start with a $(U_q \otimes U_q)$–module. The obvious way to make $M \otimes N$ into a U_q–module is to introduce a homomorphism of k–algebras $\Delta : U_q \to U_q \otimes U_q$ and let any $u \in U_q$ act on $M \otimes N$ as $\Delta(u)$ does. A simple calculation with the defining relations shows (cf. [LQG], 3.1) that we can define Δ by

$$\Delta(E) = E \otimes 1 + K \otimes E, \quad \Delta(F) = F \otimes K^{-1} + 1 \otimes F, \quad \Delta(K) = K \otimes K. \quad (1)$$

The dual space M^* of a (left) U_q–module is to start with a right U_q–module (via $(fu)(m) = f(um)$ for any $f \in M^*$, $u \in U_q$, $m \in M$). In order to get a structure as a left U_q–module, we want to have a k–linear map $S : U_q \to U_q$ satisfying $S(u_1u_2) = S(u_2)S(u_1)$ for all $u_1, u_2 \in U_q$, i.e., a ring anti-homomorphism. Then we get a left module structure where any $u \in U_q$ acts on the left as $S(u)$ acts on the right. Such an S can be defined (cf. [LQG], 3.6) by

$$S(E) = -K^{-1}E, \qquad S(F) = -FK, \qquad S(K) = K^{-1}. \tag{2}$$

This map is actually bijective, i.e., an algebra anti-automorphism; this follows (e.g.) from $S^2(u) = K^{-1}uK$ for all $u \in U_q$.

We finally need a quantum analogue of the trivial representation of a Lie algebra, i.e., a structure on k as a U_q–module. Such a structure is given by a k–algebra homomorphism $\varepsilon : U_q \to k$. We choose it such that

$$\varepsilon(E) = 0, \qquad \varepsilon(F) = 0, \qquad \varepsilon(K) = 1. \tag{3}$$

The corresponding trival module is the one denoted by $L(0, +)$ in 1.4 that made sense also in the situation from 1.5.

We call Δ the *comultiplication* of U_q, we call S the *antipode* of U_q and ε the *counit* (or *augmentation*) of U_q.

1.7. Having introduced tensor product, dual modules, and a trivial module we would like then to have the usual properties. This will not be true of all these properties. For example, looking at the modules in 1.4/5 one can easily find U_q–modules M and N such that the map $M \otimes N \to N \otimes M$ with $m \otimes n \mapsto n \otimes m$ is *not* an isomorphism of U_q–modules. The non-symmetry of the definition 1.6(1) is responsible for this phenomenon. We shall return to it later, but want to discuss first some "positive" cases. (Proofs of the claims below can be found in [LQG], 3.2 – 3.9.)

If M_1, M_2, and M_3 are U_q–modules, then the canonical isomorphism of vector spaces $(M_1 \otimes M_2) \otimes M_3 \xrightarrow{\sim} M_1 \otimes (M_2 \otimes M_3)$ with $(m_1 \otimes m_2) \otimes m_3 \mapsto m_1 \otimes (m_2 \otimes m_3)$ is an isomorphism of U_q–modules. This follows from the commutativity of the diagram

$$\begin{array}{ccccc} & & U_q \otimes U_q & \xrightarrow{1\otimes\Delta} & U_q \otimes (U_q \otimes U_q) \\ & \Delta\nearrow & & & \\ U_q & & & & \Big\downarrow \text{can} \\ & \Delta\searrow & & & \\ & & U_q \otimes U_q & \xrightarrow{\Delta\otimes 1} & (U_q \otimes U_q) \otimes U_q \end{array} \tag{1}$$

Here can is the canonical isomorphism of vector spaces with $(a \otimes b) \otimes c \mapsto a \otimes (b \otimes c)$.

The commutativity of (1) is usually expressed as: "Δ is *coassociative*".

When we regard k as an U_q–module, then we always mean the "trivial" structure given by the augmentation ε. We get now for any U_q–module M that the isomorphisms of vector spaces $M \xrightarrow{\sim} M \otimes k$ with $m \mapsto m \otimes 1$ and $M \xrightarrow{\sim} k \otimes M$ with $m \mapsto 1 \otimes m$ are isomorphisms of U_q–modules. This follows from the commutativity of the diagrams:

$$\begin{array}{ccc} U_q & \xrightarrow{\Delta} & U_q \otimes U_q \\ {\scriptstyle \mathrm{id}}\big\downarrow & & \big\downarrow{\scriptstyle 1\otimes\varepsilon} \\ U_q & \xrightarrow{\mathrm{can}} & U_q \otimes k \end{array} \qquad\qquad \begin{array}{ccc} U_q & \xrightarrow{\Delta} & U_q \otimes U_q \\ {\scriptstyle \mathrm{id}}\big\downarrow & & \big\downarrow{\scriptstyle \varepsilon\otimes 1} \\ U_q & \xrightarrow{\mathrm{can}} & k \otimes U_q \end{array} \tag{2}$$

where can denotes the isomorphism $u \mapsto u \otimes 1$ resp. $u \mapsto 1 \otimes u$.

For any U_q–module M the map $M^* \otimes M \to k$ with $f \otimes m \mapsto f(m)$ is a homomorphism of U_q–modules. This follows from the commutativity of the diagrams (actually only the second one is needed):

$$\begin{array}{ccc} U_q & \xrightarrow{\Delta} & U_q \otimes U_q \\ {\scriptstyle \iota\circ\varepsilon}\big\downarrow & & \big\downarrow{\scriptstyle 1\otimes S} \\ U_q & \xleftarrow{m} & U_q \otimes U_q \end{array} \qquad\qquad \begin{array}{ccc} U_q & \xrightarrow{\Delta} & U_q \otimes U_q \\ {\scriptstyle \iota\circ\varepsilon}\big\downarrow & & \big\downarrow{\scriptstyle S\otimes 1} \\ U_q & \xleftarrow{m} & U_q \otimes U_q \end{array} \tag{3}$$

where m is the multiplication map $m(u \otimes u') = uu'$ and where $\iota : k \to U_q$ is the embedding $\iota(a) = a\,1$.

The commutativity of the diagrams (1) – (3) means that U_q with Δ, ε, and S is a *Hopf algebra*.

1.8. If M is a finite dimensional U_q–module, then the vector space isomorphism $\psi : M \xrightarrow{\sim} (M^*)^*$ with $\psi(m)(f) = f(m)$ is usually not an isomorphism of U_q–modules. (The problem is that S^2 is not the identity.) However, there does exist an isomorphism $\psi' : M \xrightarrow{\sim} (M^*)^*$ of U_q–modules: It is given by $\psi'(m)(f) = f(K^{-1}m)$, cf. [LQG], 3.9.

The situation is similar with tensor products: The obvious bijection $M \otimes N \to N \otimes M$ is usually not an isomorphism of U_q–modules. But at least for finite dimensional M and N and q not a root of unity (and $\mathrm{char}(k) \neq 2$) it is clear that $M \otimes N$ and $N \otimes M$ are isomorphic: Recall from 1.4 that finite dimensional modules are semisimple in this case. The multiplicity of a simple module $L(n,+)$ in a direct sum decomposition of a finite dimensional module V is equal to $\dim(V_{q^n}) - \dim(V_{q^{n+2}})$, that of $L(n,-)$ is equal to $\dim(V_{-q^n}) - \dim(V_{-q^{n+2}})$. Therefore the dimensions of the eigenspaces for K determine the isomorphism class of V. Now $M \otimes N$ and $N \otimes M$ have the same eigenspaces for K since $\Delta(K) = K \otimes K$ is symmetric; so these modules are isomorphic.

An explicit construction of such an isomorphism was carried out by Drinfel'd who worked with a somewhat different version of the quantum group. You can find a version adapted to the present set-up in [LQG], 3.14. These maps lead to solutions of the quantum Yang-Baxter equations. That is the main reason why quantum groups were introduced.

Lecture 2: The general case

In this second lecture I shall introduce the quantum version of an arbitrary semisimple complex Lie algebra of finite dimension. The general conventions from the first lecture remain in force. At the beginning of 2.2 I shall introduce some restriction on the ground field k. From 2.5 on we shall assume that q is not a root of unity.

2.1. Let $\mathfrak{g}$ be a complex semisimple finite dimensional Lie algebra. Fix a Cartan subalgebra $\mathfrak{h}$ of $\mathfrak{g}$ and a corresponding triangular decomposition $\mathfrak{g} = \mathfrak{n}^- \oplus \mathfrak{h} \oplus \mathfrak{n}^+$ (as in [Ca], Section 7, but writing $\mathfrak{n}^+$ instead of $\mathfrak{n}$). So we have chosen in the root system Φ of $\mathfrak{g}$ with respect to $\mathfrak{h}$ a positive system Φ^+ and set $\mathfrak{n}^+$ equal to the direct sum of all root subspaces $\mathfrak{g}_\alpha$ with $\alpha \in \Phi^+$, while $\mathfrak{n}^-$ is the sum of all $\mathfrak{g}_{-\alpha}$ with $\alpha \in \Phi^+$.

Denote by Π the set of simple roots in Φ with respect to the choice of Φ^+. Choose for all $\alpha \in \Pi$ root vectors $e_\alpha \in \mathfrak{g}_\alpha$ and $f_\alpha \in \mathfrak{g}_{-\alpha}$ such that their commutators $h_\alpha = [e_\alpha, f_\alpha]$ satisfy $\alpha(h_\alpha) = 2$. Pick a scalar product $(\, , \,)$ on $\sum_{\alpha \in \Phi} \mathbf{R}\alpha$ that is invariant under the Weyl group of Φ. We shall assume that $(\, , \,)$ is normalised such that $(\alpha, \alpha) = 2$ for all short roots α in any component of Φ. We have $\alpha(h_\beta) = 2(\alpha, \beta)/(\beta, \beta)$ for all $\alpha, \beta \in \Pi$; let me use the abbreviation $a_{\alpha\beta} = \beta(h_\alpha)$.

The universal enveloping algebra $U(\mathfrak{g})$ has now (by a theorem of Serre) the following presentation, cf. [Hu], 18.3: It is generated by the e_α, f_α, and h_α with $\alpha \in \Pi$ with relations

$$h_\alpha h_\beta - h_\beta h_\alpha = 0, \qquad e_\alpha f_\beta - f_\beta e_\alpha = \delta_{\alpha\beta} h_\alpha,$$
$$h_\alpha e_\beta - e_\beta h_\alpha = a_{\alpha\beta} e_\beta, \qquad h_\alpha f_\beta - f_\beta h_\alpha = -a_{\alpha\beta} f_\beta,$$

and (for all $\alpha \neq \beta$)

$$\sum_{s=0}^{1-a_{\alpha\beta}} (-1)^s \binom{1-a_{\alpha\beta}}{s} e_\alpha^{1-a_{\alpha\beta}-s} e_\beta e_\alpha^s = 0$$

and

$$\sum_{s=0}^{1-a_{\alpha\beta}} (-1)^s \binom{1-a_{\alpha\beta}}{s} f_\alpha^{1-a_{\alpha\beta}-s} f_\beta f_\alpha^s = 0.$$

2.2. We now want to define the *quantised enveloping algebra* $U_q(\mathfrak{g})$ by "deforming" the relations from 2.1 similarly to the "deformation" that produced 1.1(3)–(6) out of 1.1(2).

Let us fix a field k. In order to avoid special cases (such as in the proposition from 1.4) we assume that the characteristic of k is different from 2; if Φ has a component of type G_2 assume also that $\mathrm{char}(k) \neq 3$. Pick $q \in k$, $q \neq 0$. Set $q_\alpha = q^{(\alpha,\alpha)/2}$ for all $\alpha \in \Pi$; so we have $q_\alpha \in \{q, q^2, q^3\}$. Assume that $q_\alpha^2 \neq 1$ for all $\alpha \in \Pi$.

Define $U_q(\mathfrak{g})$ as the k–algebra with generators E_α, F_α, K_α, and K_α^{-1} (for all $\alpha \in \Pi$) and relations (for all $\alpha, \beta \in \Pi$)

$$\begin{aligned} K_\alpha K_\alpha^{-1} = 1 = K_\alpha^{-1} K_\alpha, \qquad & K_\alpha K_\beta = K_\beta K_\alpha, \\ K_\alpha E_\beta K_\alpha^{-1} &= q^{(\alpha,\beta)} E_\beta, \\ K_\alpha F_\beta K_\alpha^{-1} &= q^{-(\alpha,\beta)} F_\beta, \\ E_\alpha F_\beta - F_\beta E_\alpha &= \delta_{\alpha\beta} \frac{K_\alpha - K_\alpha^{-1}}{q_\alpha - q_\alpha^{-1}}, \end{aligned}$$

and (for $\alpha \neq \beta$)

$$\sum_{s=0}^{1-a_{\alpha\beta}} (-1)^s \begin{bmatrix} 1 - a_{\alpha\beta} \\ s \end{bmatrix}_\alpha E_\alpha^{1-a_{\alpha\beta}-s} E_\beta E_\alpha^s = 0,$$

$$\sum_{s=0}^{1-a_{\alpha\beta}} (-1)^s \begin{bmatrix} 1 - a_{\alpha\beta} \\ s \end{bmatrix}_\alpha F_\alpha^{1-a_{\alpha\beta}-s} F_\beta F_\alpha^s = 0.$$

The last two relations involve the following notation:

$$\begin{bmatrix} m \\ s \end{bmatrix}_\alpha = \frac{[m]_\alpha^!}{[s]_\alpha^! [m-s]_\alpha^!} \qquad \text{where} \qquad [s]_\alpha^! = \prod_{i=1}^{s} \frac{q_\alpha^i - q_\alpha^{-i}}{q_\alpha - q_\alpha^{-1}}. \tag{1}$$

Note that we get for $\mathfrak{g} = \mathfrak{sl}_2$ the algebra U_q from 1.1. In general, we can consider for each $\alpha \in \Pi$ the subalgebra of $U_q(\mathfrak{g})$ generated by E_α, F_α, K_α, and K_α^{-1}. These elements satisfy the relations 1.1(3)–(6) with q replaced by q_α. So we have a homomorphism from the algebra U_{q_α} (in the notations of 1.1) onto that subalgebra of $U_q(\mathfrak{g})$. This homomorphism turns out to be an isomorphism, cf. [LQG], 4.22.

2.3. Proposition: *There is on $U_q(\mathfrak{g})$ a unique Hopf algebra structure (Δ, ε, S) such that for all $\alpha \in \Pi$*

$$\begin{aligned} \Delta(E_\alpha) &= E_\alpha \otimes 1 + K_\alpha \otimes E_\alpha, & \varepsilon(E_\alpha) &= 0, & S(E_\alpha) &= -K_\alpha^{-1} E_\alpha, \\ \Delta(F_\alpha) &= F_\alpha \otimes K_\alpha^{-1} + 1 \otimes F_\alpha, & \varepsilon(F_\alpha) &= 0, & S(F_\alpha) &= -F_\alpha K_\alpha, \\ \Delta(K_\alpha) &= K_\alpha \otimes K_\alpha, & \varepsilon(K_\alpha) &= 1, & S(K_\alpha) &= K_\alpha^{-1}. \end{aligned}$$

The claim says more explicitly that there exist (clearly unique) algebra homomorphisms $\Delta : U_q(\mathfrak{g}) \to U_q(\mathfrak{g}) \otimes U_q(\mathfrak{g})$ and $\varepsilon : U_q(\mathfrak{g}) \to k$ and an algebra anti-homomorphism $S : U_q(\mathfrak{g}) \to U_q(\mathfrak{g})$ given on the generators as stated and such that the diagrams 1.7(1)–(3) commute [with U_q replaced by $U_q(\mathfrak{g})$].

The proof of the existence involves checking that the relations between the generators are preserved. (That can be unpleasant, cf. [LQG], 4.8/11.) On the other hand, the proof of the commutativity of those diagrams is easily reduced to the case considered in Lecture 1. Note that the homomorphisms $U_{q_\alpha}(\mathfrak{sl}_2) \to U_q(\mathfrak{g})$ discussed at the end of 2.2 are compatible with Δ, ε, and S, i.e., they are homomorphisms of Hopf algebras.

What I said in 1.6–1.8 on tensor products, dual modules, and the trivial module generalises to the general case, cf. [LQG], 5.3, 7.3.

2.4. The triangular decomposition $\mathfrak{g} = \mathfrak{n}^- \oplus \mathfrak{h} \oplus \mathfrak{n}^+$ of the Lie algebra leads via the PBW theorem to a decomposition of the enveloping algebra $U(\mathfrak{g}) = U(\mathfrak{n}^-) \otimes U(\mathfrak{h}) \otimes U(\mathfrak{n}^+)$ [tensor product of vector spaces, not of algebras]. We have a similar decomposition of $U_q(\mathfrak{g})$.

Before I describe it, let me simplify the notation. Keep q and $\mathfrak{g}$ fixed and write just $U = U_q(\mathfrak{g})$. Denote by U^+ (resp. U^-) the subalgebra of U generated by all E_α (resp. F_α) with $\alpha \in \Pi$. Set U^0 equal to the subalgebra of U generated by all K_α and K_α^{-1} with $\alpha \in \Pi$. We get then (cf. [LQG], 4.21.a)

Theorem: *The multiplication map*

$$U^- \otimes U^0 \otimes U^+ \longrightarrow U, \qquad u_1 \otimes u_2 \otimes u_3 \mapsto u_1 u_2 u_3 \tag{1}$$

is an isomorphism of vector spaces.

This is a partial generalisation of the Theorem in 1.2. That earlier result says not only that our present theorem holds for $\mathfrak{g} = \mathfrak{sl}_2$, but also that (in that case) the E^r are a basis of U^+, the F^s a basis for U^-, and the K^n a basis for U^0.

In order to generalise the Theorem in 1.2 completely, we should have explicit bases for U^-, U^0, and U^+ in general. Then the theorem above says that we get a basis for U by taking products [as in (1)] of these basis elements. Finding such a basis is easy for U^0. For each $\mu = \sum_{\alpha\in\Pi} m(\alpha)\alpha$ in the root lattice $\mathbf{Z}\Phi$ set $K_\mu = \prod_{\alpha\in\Pi} K_\alpha^{m(\alpha)}$. Then, cf. [LQG], 4.21.d:

Proposition: *The K_μ with $\mu \in \mathbf{Z}\Phi$ are a basis of U^0.*

Finding bases for U^+ and U^- is somewhat more complicated in the general case and we will discuss bases only in Lecture 3.

2.5. Assume from now on that q in *not* a root of unity.

We now want to describe finite dimensional U–modules. Let me start with looking at one dimensional representations. These correspond to algebra homomorphisms $U \to k$. Looking at the relations in 2.2 one sees easily that all E_α and F_α have to go to 0 while each K_α has to go to 1 or -1. Conversely, any choice of signs for the images of the K_α yields a homomorphism.

Set $\mathfrak{S}$ equal to the set of all maps from Π to $\{1,-1\} \subset k$. This set contains $2^{|\Pi|}$ elements since $\mathrm{char}(k) \neq 2$. The result above can be expressed as follows: We get for each $\sigma \in \mathfrak{S}$ a one dimensional U–module k_σ, where all E_α and F_α act as 0 and where any K_α acts as $\sigma(\alpha)$. We get $2^{|\Pi|}$ non-isomorphic one dimensional U–modules, and each one dimensional U–module is isomorphic to one of these k_σ.

Denote by X the weight lattice of the root system Φ, i.e., the set of all $\lambda \in \mathbf{Q}\Phi$ with $\frac{2(\lambda,\alpha)}{(\alpha,\alpha)} \in \mathbf{Z}$ for all $\alpha \in \Pi$. Our normalisation of (,) implies that $(\lambda, \alpha) \in \mathbf{Z}$ for all $\lambda \in X$ and $\alpha \in \Pi$. So $q^{(\lambda,\alpha)}$ makes sense for all these λ and σ. Given a U–module V, set

$$V_{\lambda,\sigma} = \{\, v \in V \mid K_\alpha v = \sigma(\alpha)\, q^{(\lambda,\alpha)}\, v \quad \text{for all } \alpha \in \Pi \,\} \tag{1}$$

for all $\lambda \in X$ and $\sigma \in \mathfrak{S}$.

The $V_{\lambda,\sigma}$ are subspaces of V called *weight spaces*. Since q is not a root of unity, distinct λ and σ lead to distinct $V_{\lambda,\sigma}$; it follows that the sum of the $V_{\lambda,\sigma}$ is direct. If V is finite dimensional, then we get

$$V = \bigoplus_{\lambda \in X} \bigoplus_{\sigma \in \mathfrak{S}} V_{\lambda,\sigma}. \tag{2}$$

This follows from the results in 1.4 applied for each $\alpha \in \Pi$ to the subalgebra generated by E_α, F_α, K_α, and K_α^{-1}. It shows that the action of each K_α on V is diagonalisable with each eigenvalue of the form $\pm q^m$ with $m \in \mathbf{Z}$. Since the K_α commute, they can be diagonalised simultaneously. One checks then, that each common eigenspace is one of the $V_{\lambda,\sigma}$.

2.6. We are now ready to state the generalisations of the results from 1.4. To start with, we have (cf. [LQG], 5.17)

Proposition 1: *Every finite dimensional U–module is semisimple.*

Next we have a classification of the simple modules: We call a U–module V a *highest weight module* with highest weight (λ, σ) if $\dim(V_{\lambda,\sigma}) = 1$, if $V_{\lambda,\sigma}$ generates V as a U–module, and if $E_\alpha V_{\lambda,\sigma} = 0$ for all $\alpha \in \Pi$. Recall that a weight $\lambda \in X$ is called *dominant* if $(\lambda, \alpha) \geq 0$ for all $\alpha \in \Pi$. We have now (cf. [LQG], 5.10 and 5.2):

Proposition 2: *For each dominant $\lambda \in X$ and each $\sigma \in \mathfrak{S}$ there exists a finite dimensional and simple U–module $L(\lambda, \sigma)$ that is a highest weight* *module with highest weight (λ, σ). Each finite dimensional simple U–module is isomorphic to exactly one of these $L(\lambda, \sigma)$.*

It is easy to see (cf. [LQG], 5.2) that any $L(\lambda, \sigma)$ is the direct sum of all $L(\lambda, \sigma)_{\mu,\sigma}$, i.e., that $L(\lambda, \sigma)_{\mu,\sigma'} = 0$ whenever $\sigma' \neq \sigma$. The dimensions of the non-zero weight spaces are given by Weyl's character formula. This follows from (cf. [LQG], 5.15 and 6.26):

Proposition 3: *Let $\sigma \in \mathfrak{S}$ and $\lambda, \mu \in X$ with λ dominant. Then $L(\lambda, \sigma)_{\mu,\sigma}$ has the same dimension as the μ weight space of the simple $\mathfrak{g}$–module with highest weight λ.*

One final observation: Write $L(\lambda)$ for the $L(\lambda, \sigma)$ with $\sigma(\alpha) = 1$ for all α. Then we get for arbitrary σ that

$$L(\lambda, \sigma) \simeq L(\lambda) \otimes k_\sigma. \tag{1}$$

2.7. Let me briefly describe the centre $Z(U)$ of U and generalise Proposition 1.3.a. We first construct an analogue of the Harish-Chandra homomorphism used to describe the centre of $U(\mathfrak{g})$, cf. [Dix], 7.4.

Set $U^{++} = \ker(\varepsilon) \cap U^+$ and $U^{--} = \ker(\varepsilon) \cap U^-$. We have then $U^+ = k1 \oplus U^{++}$ and $U^- = k1 \oplus U^{--}$, hence $U = U^0 \oplus (U^{--}U^0U^+ + U^-U^0U^{++})$ by the theorem in 2.4. Denote by $\pi : U \to U^0$ the projection with respect to this decomposition. The restriction of π to $Z(U)$ is an algebra homomorphism, cf. [LQG], 6.2.

There is for each $\lambda \in X$ a unique algebra automorphism γ_λ of U^0 with $\gamma_\lambda(K_\alpha) = q^{(\lambda,\alpha)} K_\alpha$ for all $\alpha \in \Pi$. Set ρ equal to half the sum of all positive roots in Φ. This is an element of X, so $\gamma_{-\rho}$ makes sense. We call now $\gamma_{-\rho} \circ \pi : Z(U) \to U^0$ the *Harish-Chandra homomorphism* of U. The first result says now (cf. [LQG], 6.3.b):

Proposition 1: *The Harish-Chandra homomorphism of U is injective.*

In order to describe the image of this map we need the Weyl group W of our root system Φ. It permutes Φ, hence $\mathbf{Z}\Phi$. We get an action of W on U^0 via $w(K_\mu) = K_{w\mu}$ for all $\mu \in \mathbf{Z}\Phi$ and $w \in W$. Set U^0_{ev} equal to the span of all K_μ with $\mu \in \mathbf{Z}\Phi \cap 2X$. This is a subalgebra of U^0 stable under the action of W. The quantum analogue of Harish-Chandra's classical description of the centre of $U(\mathfrak{g})$ is now (cf. [LQG], 6.25/26):

Proposition 2: *The Harish-Chandra homomorphism has image $(U^0_{ev})^W$.*

Combining these propositions we see that the Harish-Chandra homomorphism is an isomorphism $Z(U) \xrightarrow{\sim} (U^0_{ev})^W$.

2.8. Suppose here that k has characteristic 0, that q is transcendental over $\mathbf{Q}$, and that $k = \mathbf{Q}(q)$. Set $A = \mathbf{Z}[q, q^{-1}] \subset k$. So A is a subring of k; it has field of fractions equal to k and it is a free $\mathbf{Z}$–module with basis all q^n with $n \in \mathbf{Z}$.

We are going to describe a quantum analogue of the "Kostant $\mathbf{Z}$–form" $U(\mathfrak{g})_{\mathbf{Z}}$ of $U(\mathfrak{g})$, cf. [Hu], 26.4. That $\mathbf{Z}$–form is the $\mathbf{Z}$–subalgebra of $U(\mathfrak{g})$ generated by all $x_\alpha^n/n!$ with $\alpha \in \Phi$ and $n \in \mathbf{Z}$, $n \geq 0$ where the x_α with $\alpha \in \Phi$ are part of a Chevalley basis of $\mathfrak{g}$, cf. [Hu], 25.2. One can show that $U(\mathfrak{g})_{\mathbf{Z}}$ is generated already by the $x_\alpha^n/n!$ with $\alpha \in (\Pi \cup -\Pi)$.

Set U_A equal to the A–subalgebra of U generated by all $E_\alpha^n/[n]_\alpha^!$ and $F_\alpha^n/[n]_\alpha^!$ with $\alpha \in \Pi$ and $n \in \mathbf{Z}$, $n \geq 0$, and by all K_α and K_α^{-1}. (Recall the definition of $[n]_\alpha^!$ in 2.2(1).)

It turns out that U_A is a free A–module; any basis of U_A over A is also a basis of U over k. There is an analogue of the triangular decomposition from 2.4: Multiplication induces an isomorphism of A–modules $U_A^- \otimes_A U_A^0 \otimes_A U_A^+ \simeq U_A$ where $U_A^- = U_A \cap U^-$, etc. (See [Lu1], Thm. 6.7.)

Given a field K and a non-zero element $v \in K$, we have a unique ring homomorphism $A = \mathbf{Z}[q, q^{-1}] \to K$ with $q \mapsto v$. This makes K into an A–algebra and we can construct

$$U_{K,v} = U_A \otimes_A K. \tag{1}$$

If v is not a root of unity, then this new algebra is (isomorphic to) the quantised enveloping algebra $U_v(\mathfrak{g})$ constructed over K. To start with, the elements $E_\alpha \otimes 1$, $F_\alpha \otimes 1$, and $K_\alpha^{\pm 1} \otimes 1$ in $U_{K,v}$ satisfy the defining relations of $U_v(\mathfrak{g})$. We get thus a natural homomorphism from $U_v(\mathfrak{g})$ to $U_{K,v}$. Using that v is not a root of unity one checks next that each $[n]_\alpha^!$ (which is in A) has a non-zero image in K. This implies that $U_{K,v}$ is generated by all $E_\alpha \otimes 1$, $F_\alpha \otimes 1$, and $K_\alpha^{\pm 1} \otimes 1$, hence that our homomorphism $U_v(\mathfrak{g}) \to U_{K,v}$ is surjective. The injectivity follows then comparing dimensions of weight spaces in U^+ and in U^- and by an explicit look at bases for U^0.

This discussion shows that we can get something new only when v is a root of unity. Let us consider the extreme case: $v = 1$. Set $\mathfrak{g}_{\mathbf{Z}}$ equal to the $\mathbf{Z}$–span of a Chevalley basis of $\mathfrak{g}$ and set $\mathfrak{g}_K = \mathfrak{g}_{\mathbf{Z}} \otimes_{\mathbf{Z}} K$; so this is a split Lie algebra over K of the same type as $\mathfrak{g}$. If K has characteristic 0, then we have (cf. [Lu1], 8.16) a surjective homomorphism from $U_{K,1}$ to the enveloping algebra $U(\mathfrak{g}_K)$. The kernel of this map is the ideal generated by all $K_\alpha \otimes 1 - 1$; the $K_\alpha \otimes 1$ are central in $U_{K,1}$ and satisfy $(K_\alpha \otimes 1)^2 = 1$. This is why we (loosely) think of $U_q(\mathfrak{g})$ as a deformation of $U(\mathfrak{g})$. More rigorously, it is a deformation of a "central" extension of $U(\mathfrak{g})$.

If we take K of characteristic $p > 0$, then $U_{K,1}$ maps not onto $U(\mathfrak{g}_K)$, but onto $U(\mathfrak{g})_{\mathbf{Z}} \otimes_{\mathbf{Z}} K$, the reduction modulo p of the "Kostant $\mathbf{Z}$–form" of $U(\mathfrak{g})$. The kernel of this surjection has a description similar to the one in the characteristic 0 case.

Consider finally the case where v is a root of unity such that $U_v(\mathfrak{g})$ is defined, i.e., such that v satisfies the condition on q at the beginning of

2.2. We have then (as in the first case discussed) a natural homomorphism $U_v(\mathfrak{g}) \to U_{K,v}$, but now that map is far from being surjective. Instead, its image is finite dimensional. The relation between $U_v(\mathfrak{g})$ and $U_{K,v}$ is similar to that between $U(\mathfrak{g}_F)$ and $U(\mathfrak{g})_{\mathbf{Z}} \otimes_{\mathbf{Z}} F$ for a field F of characteristic $p > 0$. There, too, is a natural homomorphism $U(\mathfrak{g}_F) \to U(\mathfrak{g})_{\mathbf{Z}} \otimes_{\mathbf{Z}} F$ with a finite dimensional image, isomorphic to the restricted enveloping algebra of the restricted Lie algebra $\mathfrak{g}_F$. So the image of the homomorphism $U_v(\mathfrak{g}) \to U_{K,v}$ is a quantum analogue to that restricted enveloping algebra.

Lecture 3: Bases

Keep the notations and assumptions from Lecture 2. In 3.1–3.5 we assume that q is not a root of unity. (This assumption could be avoided; cf. the discussion in [AJS], 1.3.) From 3.7 on we assume that k and q are as in 2.8.

3.1. If we want to find a basis for $U = U_q(\mathfrak{g})$, it suffices to find bases for U^+ and U^- (as pointed out in 2.4). A look at the defining relations of U shows that there exists an automorphisms ω of U with $\omega(E_\alpha) = F_\alpha$, $\omega(F_\alpha) = E_\alpha$, and $\omega(K_\alpha) = K_\alpha^{-1}$ for all $\alpha \in \Pi$. It satisfies $\omega^2 = 1$ and maps U^+ isomorphically to U^-. The image of a basis of U^+ is a basis for U^- (and vice versa). So it suffices to find a basis for one of these two algebras.

3.2. The algebra U^+ is (in some sense) a deformation of the enveloping algebra $U(\mathfrak{n}^+)$. So we should expect them to have similar bases. The PBW theorem says that we get a basis of $U(\mathfrak{n}^+)$ as follows: We fix a numbering $\gamma_1, \gamma_1, \dots, \gamma_N$ of the positive roots and we choose for each i a root vector $e_{\gamma_i} \in \mathfrak{g}_{\gamma_i}$, $e_{\gamma_i} \neq 0$. Then all products $e_{\gamma_1}^{m(1)} e_{\gamma_2}^{m(2)} \dots e_{\gamma_N}^{m(N)}$ with all $m(i) \geq 0$ are a basis of $U(\mathfrak{n}^+)$.

In order to imitate this construction for U^+, we need analogues to the root vectors e_γ. For a simple root α the obvious choice is the generator E_α. But in general there is not a clear candidate.

Let us see how to get (in the Lie algebra case) the e_γ for all γ when we have them at first only for simple roots: Given an arbitrary root γ we can find a simple root β and an element w in the Weyl group W with $\gamma = w(\beta)$. Then we take a connected algebraic group G with Lie algebra $\mathfrak{g}$ and a maximal torus T in G with Lie algebra $\mathfrak{h}$. Then W can be identified with $N_G(T)/T$; we pick a representative $\dot{w} \in N_G(T)$ of w and can then choose $e_\gamma = \operatorname{Ad}(\dot{w}) e_\beta$ where Ad denotes the adjoint action of G on its Lie algebra.

If we want to do something similar for U, then we need a quantum analogue for the adjoint action of $\dot{w}$. Since W is generated by the reflections s_α with $\alpha \in \Pi$, it suffices to construct these operators for $w = s_\alpha$. There are several reasonable possibilities for these operators; here is one (cf. [LQG], 8.13/14):

Theorem: *Let $\alpha \in \Pi$. Then there exists an automorphism T_α of U such that*

$$T_\alpha(K_\mu) = K_{s_\alpha \mu} \qquad \text{for all } \mu \in \mathbf{Z}\Phi, \tag{1}$$

and

$$T_\alpha(E_\alpha) = -F_\alpha K_\alpha, \qquad T_\alpha(F_\alpha) = -K_\alpha^{-1} E_\alpha, \tag{2}$$

and for all $\beta \in \Pi$, $\beta \neq \alpha$

$$T_\alpha(E_\beta) = \sum_{i=0}^{-a_{\alpha\beta}} (-1)^i q_\alpha^{-i} E_\alpha^{(-a_{\alpha\beta}-i)} E_\beta E_\alpha^{(i)}, \tag{3}$$

$$T_\alpha(F_\beta) = \sum_{i=0}^{-a_{\alpha\beta}} (-1)^i q_\alpha^{i} F_\alpha^{(i)} F_\beta F_\alpha^{(-a_{\alpha\beta}-i)}, \tag{4}$$

where $a_{\alpha\beta} = \beta(h_\beta)$ *as in* 2.1.

We use here the notations

$$E_\alpha^{(n)} = \frac{E_\alpha^n}{[n]_\alpha^!} \qquad \text{and} \qquad F_\alpha^{(n)} = \frac{F_\alpha^n}{[n]_\alpha^!} \tag{5}$$

for all integers $n \geq 0$ with $[n]_\alpha^! \neq 0$. If q is not a root of unity, then a look at 2.2(1) shows that $[n]_\alpha^! \neq 0$ for all $n \geq 0$. The restrictions on q in 2.2 make sure that $[n]_\alpha^! \neq 0$ for all $n \leq |a_{\alpha\beta}|$ for all $\alpha \neq \beta$. Therefore (3) and (4) make sense in all cases.

The simplest case is where $\beta(h_\alpha) = 0$; in that situation (3) and (4) say

$$T_\alpha(E_\beta) = E_\beta \qquad \text{and} \qquad T_\alpha(F_\beta) = F_\beta. \tag{6}$$

If $\beta(h_\alpha) = -1$ we get

$$T_\alpha(E_\beta) = E_\alpha E_\beta - q_\alpha^{-1} E_\beta E_\alpha \qquad \text{and} \qquad T_\alpha(F_\beta) = F_\beta F_\alpha - q_\alpha^{-1} F_\alpha F_\beta. \tag{7}$$

3.3. The first main property of these T_α is (cf. [LQG], 8.15–17):

Theorem: *The* T_α *with* $\alpha \in \Pi$ *satisfy the braid relations.*

This says more explicitly (cf. [Ro]) that we have for all $\alpha \neq \beta$ in Π

$$T_\alpha T_\beta \cdots = T_\beta T_\alpha \cdots \tag{1}$$

where the number of factors on both sides is equal to the order of $s_\alpha s_\beta$ in W.

The theorem says that we have an action of the *braid group* B_W on U. Actually more is true: If q is not a root of unity, then we have an action of B_W on each finite dimensional U–module V that is compatible with the action of B_W on U, i.e., with $T_\alpha(uv) = T_\alpha(u)T_\alpha(v)$ for all $u \in U$ and $v \in V$, cf. [Lu2], ch. 39.

At this point we need just one consequence of the braid relations: There is for each $w \in W$ an automorphism T_w of U such that whenever $w =$

$s_{\alpha_1} s_{\alpha_2} \dots s_{\alpha_r}$ with all $\alpha_i \in \Pi$ and with r minimal (i.e., with r equal to the length of w as in [Ro], before Thm. 2.3), then

$$T_w = T_{\alpha_1} T_{\alpha_2} \cdots T_{\alpha_r}. \tag{1}$$

3.4. By analogy with the Lie algebra case we want to define $E_\gamma = T_w(E_\beta)$ if $w \in W$ and $\beta \in \Pi$ with $\gamma = w(\beta)$. If we want to use this element to construct a basis of U^+, it should be contained in U^+. That turns out to be true (cf. [LQG], 8.20):

Proposition: *Let $w \in W$ and $\alpha \in \Pi$. If $w\alpha > 0$, then $T_w(E_\alpha) \in U^+$; if $w\alpha \in \Pi$, then $T_w(E_\alpha) = E_{w\alpha}$.*

3.5. This is nice, but does not make all problems disappear. The main problem is that the freedom of choosing $w \in W$ and $\beta \in \Pi$ with $\gamma = w(\beta)$ can lead to fundamentally distinct candidates for E_γ. Take as an example Φ of type A_2 with $\Pi = \{\alpha, \beta\}$; consider $\gamma = \alpha + \beta$. We have then $\gamma = s_\alpha(\beta)$ and $\gamma = s_\beta(\alpha)$. Therefore both $T_{s_\alpha}(E_\beta) = E_\alpha E_\beta - q^{-1} E_\beta E_\alpha$ and $T_{s_\beta}(E_\alpha) = E_\beta E_\alpha - q^{-1} E_\alpha E_\beta$ are natural choices for E_γ. However, these two vectors are linearly independent. (In the Lie algebra case different choices in the construction may produce different vectors, but they are all proportional.)

It turns out that one can make a systematic choice of root vectors as follows: Take the longest element $w_0 \in W$, i.e., the element with $w_0(\Pi) = -\Pi$. Choose a "reduced expression" $w_0 = s_{\alpha_1} s_{\alpha_2} \dots s_{\alpha_N}$ with all $\alpha_i \in \Pi$ and with $N = |\Phi^+|$. Set $\gamma_i = s_{\alpha_1} \dots s_{\alpha_{i-1}} \alpha_i$ for $1 \le i \le t$. Then $\gamma_1, \gamma_2, \dots, \gamma_t$ are exactly the positive roots in Φ each occurring once, cf. [Bou], Chap. VI, §1, Cor. 2 de la Prop. 17. Set $w_i = s_{\alpha_1} s_{\alpha_2} \dots s_{\alpha_{i-1}}$ and $E_{\gamma_i} = T_{w_i} E_{\alpha_i}$ for all i. In case $\gamma_i \in \Pi$ Proposition 3.4 says that this notation is compatible with the earlier one; in general we get from Proposition 3.4 that $E_{\gamma_i} \in U^+$. We get now (cf. [LQG], 8.24)

Theorem: *All products*

$$E_{\gamma_N}^{m(N)} E_{\gamma_{N-1}}^{m(N-1)} \dots E_{\gamma_2}^{m(2)} E_{\gamma_1}^{m(1)} \tag{1}$$

with all $m(i) \in \mathbf{Z}$, $m(i) \ge 0$ are a basis of U^+.

Remark: Consider the situation from 2.8 with $k = \mathbf{Q}(q)$. Set again $A = \mathbf{Z}[q, q^{-1}]$ and U_A, etc. as there. Then one gets (cf. the appendix to [Lu1]) that all products

$$E_{\gamma_N}^{(m(N))} E_{\gamma_{N-1}}^{(m(N-1))} \dots E_{\gamma_2}^{(m(2))} E_{\gamma_1}^{(m(1))} \tag{2}$$

with the $m(i)$ as above are a basis of U_A^+ over A, where we use the notation $E_{\gamma_i}^{(r)} = T_{w_i} E_{\alpha_i}^{(r)}$ for all $r \ge 0$.

3.6. So far we have mainly seen that quantum groups share many properties with enveloping algebras. In the second half of this lecture I want to describe an example where quantum groups have led to a new important result on enveloping algebras.

For each dominant weight $\lambda \in X$ denote the simple $\mathfrak{g}$–module with highest weight λ by $V(\lambda)$. Choose a non-zero weight vector $v_\lambda \in V(\lambda)$ of weight λ; we have then $V(\lambda) = U(\mathfrak{n}^-)v_\lambda$ cf. [Hu], 20.2. Now the new result is (cf. [LQG], 11.19):

Theorem: *There is a basis* $\mathbf{B}$ *of* $U(\mathfrak{n}^-)$ *such that for all dominant weights* λ *the* uv_λ *with* $u \in \mathbf{B}$ *and* $uv_\lambda \neq 0$ *are a basis of* $V(\lambda)$.

The point here is that one can find one basis that works simultaneously for *all* λ.

3.7. There are several approaches to Theorem 3.6. I shall describe here Kashiwara's that is somewhat more elementary than Lusztig's (to be found in [Lu2]).

Assume from now on that we are in the situation of 2.8, i.e., that $k = \mathbf{Q}(q)$ with q transcendental over $\mathbf{Q}$. We set again $A = \mathbf{Z}[q, q^{-1}]$ and introduce the local ring $R = \{\frac{f}{g} \mid f, g \in \mathbf{Z}[q], g(0) \neq 0\}$. So A and R are subrings of k with $A \cap R = \mathbf{Z}[q]$; we have natural isomorphisms $R/qR \simeq \mathbf{Q}$ and $(A \cap R)/q(A \cap R) \simeq \mathbf{Z}$.

Consider (as in 2.6) for each dominant weight λ the simple U–module $L(\lambda)$ with highest weight (λ, σ) where $\sigma(\alpha) = 1$ for all α. We shall write $L(\lambda)_\mu$ instead of $L(\lambda)_{\mu,\sigma}$ for the weight spaces of this module. Choose a weight vector $v_\lambda \in L(\lambda)_\lambda$, $v_\lambda \neq 0$; we have then $L(\lambda) = U^- v_\lambda$.

Let $\alpha \in \Pi$. For any $\mu \in X$ the (direct) sum of all $L(\lambda)_{\mu+i\alpha}$ with $i \in \mathbf{Z}$ is a submodule of $L(\lambda)$ for the subalgebra of U generated by E_α, F_α, K_α, and K_α^{-1}. This subalgebra is isomorphic to $U_{q_\alpha}(\mathfrak{sl}_2)$; we can therefore apply the representation theory of $U_q(\mathfrak{sl}_2)$ as developed in 1.4. It implies that we can write any $v \in L(\lambda)_\mu$ as a sum $v = \sum_{i \geq 0} F_\alpha^{(i)} v_i$ with $v_i \in L(\lambda)_{\mu+i\alpha}$ and $E_\alpha v_i = 0$ (and with $F_\alpha^{(i)}$ as in 3.2(5)). We assume that $v_i = 0$ whenever $F_\alpha^{(i)} v_i = 0$; then the v_i are uniquely determined by v and we can define

$$\widetilde{F}_\alpha v = \sum_{i \geq 0} F_\alpha^{(i+1)} v_i \qquad \text{and} \qquad \widetilde{E}_\alpha v = \sum_{i > 0} F_\alpha^{(i-1)} v_i. \tag{1}$$

Extend $\widetilde{F}_\alpha$ and $\widetilde{E}_\alpha$ linearly to $L(\lambda)$.

Set $\mathcal{L}(\lambda) \subset L(\lambda)$ equal to span over R of all $\widetilde{F}_{\alpha_1} \widetilde{F}_{\alpha_2} \dots \widetilde{F}_{\alpha_r} v_\lambda$ for all integers $r \geq 0$ and all series $\alpha_1, \alpha_2, \dots, \alpha_r$ of simple roots. Set $\mathcal{B}(\lambda) \subset \mathcal{L}(\lambda)/q\mathcal{L}(\lambda)$ equal to the set of all *nonzero* cosets modulo $q\mathcal{L}(\lambda)$ of all these $\widetilde{F}_{\alpha_1} \widetilde{F}_{\alpha_2} \dots \widetilde{F}_{\alpha_r} v_\lambda$. Then (cf. [LQG], 9.25):

Theorem: a) *$\mathcal{L}(\lambda)$ is a finitely generated R–module; it spans $L(\lambda)$ over k.*

b) $\mathcal{B}(\lambda)$ *is a basis of the vector space* $\mathcal{L}(\lambda)/q\mathcal{L}(\lambda)$ *over* $\mathbf{Q}$.

c) *We have* $\tilde{F}_\alpha\mathcal{L}(\lambda) \subset \mathcal{L}(\lambda)$ *and* $\tilde{E}_\alpha\mathcal{L}(\lambda) \subset \mathcal{L}(\lambda)$ *for all* $\alpha \in \Pi$. *The induced maps on* $\mathcal{L}(\lambda)/q\mathcal{L}(\lambda)$ *satisfy* $\tilde{F}_\alpha\mathcal{B}(\lambda) \subset \mathcal{B}(\lambda) \cup \{0\}$ *and* $\tilde{E}_\alpha\mathcal{B}(\lambda) \subset \mathcal{B}(\lambda) \cup \{0\}$.

d) *We have for all* $b, b' \in \mathcal{B}(\lambda)$ *and* $\alpha \in \Pi$: $\quad b = \tilde{E}_\alpha b' \iff b' = \tilde{F}_\alpha b.$

3.8. Let me insert some remarks before I continue with the construction. Working in $\mathcal{L}(\lambda)/q\mathcal{L}(\lambda)$ means that we somehow work at $q = 0$ (which of course is impossible in the original construction of $U_q(\mathfrak{g})$). Thinking of q as a temperature with $q = 0$ as the minimum, Kashiwara has coined the term *crystal base* for the pair $(\mathcal{B}(\lambda), \mathcal{L}(\lambda))$. By abuse of notation I shall use the term 'crystal basis' also for other related bases.

For $\mathfrak{sl}_2$ and for $U_q(\mathfrak{sl}_2)$ the simple modules have nice bases that are permuted (up to scalars) by the basis elements e and f (for $\mathfrak{sl}_2$) or E and F (for $U_q(\mathfrak{sl}_2)$). For general $\mathfrak{g}$ such bases do not exist. Given (e.g.) a simple $U_q(\mathfrak{g})$ module V and a simple root α, we can find a basis that is permuted (up to scalars) by E_α and F_α, but then the E_β and F_β with $\beta \neq \alpha$ will not have a nice form with respect to this basis.

Now Theorem 3.6.c says that "at $q = 0$" a basis exists such that *all* E_α and F_α (or rather their modifications $\tilde{E}_\alpha$ and $\tilde{F}_\alpha$) look nicely with respect to this basis.

Another important feature of these crystal bases is their compatibility with tensor products. Let λ and λ' be two dominant weights. Then we can find a decomposition $L(\lambda) \otimes L(\lambda') = \bigoplus_i L(\mu_i)$ such that $\mathcal{L}(\lambda) \otimes \mathcal{L}(\lambda')$ is the direct sum of all $\mathcal{L}(\mu_i)$ and such that $\mathcal{B}(\lambda) \otimes \mathcal{B}(\lambda')$ is the disjoint sum of all $\mathcal{B}(\mu_i)$ (cf. [LQG], 9.17). Furthermore, the action of the $\tilde{E}_\alpha$ and $\tilde{F}_\alpha$ on $\mathcal{B}(\lambda) \otimes \mathcal{B}(\lambda')$ can be read off their actions on $\mathcal{B}(\lambda)$ and $\mathcal{B}(\lambda')$. This can be used to determine the decomposition of $L(\lambda) \otimes L(\lambda')$ into irreducibles, cf. [LQG], 9.27.

3.9. We now want to imitate the construction from 3.7 working with U^- instead of the $L(\lambda)$. The main thing to be changed is the definition of the $\tilde{E}_\alpha$ and $\tilde{F}_\alpha$, since we do not have a $U_q(\mathfrak{sl}_2)$–action on U^-.

Let $\alpha \in \Pi$. There exist for each $u \in U^-$ elements $r_\alpha(u)$ and $r'_\alpha(u)$ in U^- with

$$E_\alpha u - u E_\alpha = \frac{K_\alpha r_\alpha(u) - r'_\alpha(u) K_\alpha^{-1}}{q_\alpha - q_\alpha^{-1}}, \tag{1}$$

cf. [LQG], 10.1(1). (This holds for the generators F_β by the defining relations in 2.2 and follows in general by induction.) One shows next that there are for each $u \in U^-$ uniquely determined elements $u_n \in U^-$, almost all equal to 0, with $r'_\alpha(u_n) = 0$ for all n, such that $u = \sum_{n\geq 0} F_\alpha^{(n)} u_n$, cf. [LQG], 10.2. We define now

$$\tilde{F}_\alpha u = \sum_{n\geq 0} F_\alpha^{(n+1)} u_n \qquad \text{and} \qquad \tilde{E}_\alpha u = \sum_{n>0} F_\alpha^{(n-1)} u_n. \tag{2}$$

Set $\mathcal{L}(\infty) \subset U^-$ equal to the span over R of all $\widetilde{F}_{\alpha_1}\widetilde{F}_{\alpha_2}\ldots\widetilde{F}_{\alpha_r}1$ with $r \geq 0$ and $\alpha_1, \alpha_2, \ldots, \alpha_r \in \Pi$. We denote by $\mathcal{B}(\infty) \subset \mathcal{L}(\infty)/q\mathcal{L}(\infty)$ the set of all cosets of these $\widetilde{F}_{\alpha_1}\widetilde{F}_{\alpha_2}\ldots\widetilde{F}_{\alpha_r}1$. We get analogously to Theorem 3.7 (cf. [LQG], 10.10–12)

Theorem: a) *$\mathcal{L}(\infty)$ is a free R–module; it generates U^- over k.*

b) *$\mathcal{B}(\infty)$ is a basis of the vector space $\mathcal{L}(\infty)/q\mathcal{L}(\infty)$ over $\mathbf{Q}$.*

c) *We have $\widetilde{F}_\alpha\mathcal{L}(\infty) \subset \mathcal{L}(\infty)$ and $\widetilde{E}_\alpha\mathcal{L}(\infty) \subset \mathcal{L}(\infty)$ for all $\alpha \in \Pi$. The induced maps on $\mathcal{L}(\infty)/q\mathcal{L}(\infty)$ satisfy $\widetilde{F}_\alpha\mathcal{B}(\infty) \subset \mathcal{B}(\infty)$ and $\widetilde{E}_\alpha\mathcal{B}(\infty) \subset \mathcal{B}(\infty) \cup \{0\}$.*

d) *Let $\alpha \in \Pi$. We have $\widetilde{E}_\alpha\widetilde{F}_\alpha b = b$ for all $b \in \mathcal{B}(\infty)$ and $\widetilde{F}_\alpha\widetilde{E}_\alpha b = b$ whenever $\widetilde{E}_\alpha b \neq 0$.*

3.10. We can now relate the constructions from 3.7 and 3.9. We have for each dominant $\lambda \in X$ a surjection $\varphi_\lambda : U^- \to L(\lambda)$ with $\varphi_\lambda(u) = uv_\lambda$. We get now first (cf. [LQG], 10.10):

Proposition 1: *We have $\varphi_\lambda\mathcal{L}(\infty) = \mathcal{L}(\lambda)$ for all dominant λ.*

It follows that φ_λ induces a linear map $\overline{\varphi}_\lambda : \mathcal{L}(\infty)/q\mathcal{L}(\infty) \to \mathcal{L}(\lambda)/q\mathcal{L}(\lambda)$. We can apply this map to elements in $\mathcal{B}(\infty)$ and get (cf. [LQG], 10.14):

Proposition 2: *For any dominant λ the map $b \mapsto \overline{\varphi}_\lambda(b)$ induces a bijection*

$$\{\, b \in \mathcal{B}(\infty) \mid \overline{\varphi}_\lambda(b) \neq 0 \,\} \xrightarrow{\sim} \mathcal{B}(\lambda).$$

3.11. The last proposition says that the basis $\mathcal{B}(\infty)$ "at $q = 0$" has the property that $\mathbf{B}$ in Theorem 3.6 is said to have "at $q = 1$". In order to get from $q = 0$ to $q = 1$ we first want to lift to general q.

Recall the construction of U_A and U_A^- from 2.8. Set $\mathcal{L}_{\mathbf{Z}}(\infty) = U_A^- \cap \mathcal{L}(\infty)$. This is a $\mathbf{Z}[q]$–module since $\mathbf{Z}[q] = A \cap R$. We have $q\mathcal{L}_{\mathbf{Z}}(\infty) = U_A^- \cap q\mathcal{L}(\infty)$ since q is a unit in A. This implies that the $\mathbf{Z}$–module $\mathcal{L}_{\mathbf{Z}}(\infty)/q\mathcal{L}_{\mathbf{Z}}(\infty)$ can be identified with the image of $\mathcal{L}_{\mathbf{Z}}(\infty)$ in the $\mathbf{Q}$–vector space $\mathcal{L}(\infty)/q\mathcal{L}(\infty)$. One can show (cf. [LQG], 11.6(5)) that under this identification

$$\mathcal{B}(\infty) \subset \mathcal{L}_{\mathbf{Z}}(\infty)/q\mathcal{L}_{\mathbf{Z}}(\infty) \subset \mathcal{L}(\infty)/q\mathcal{L}(\infty). \tag{1}$$

Using the defining relations of U one checks easily that there is a (clearly unique) automorphism ψ of U regarded as a $\mathbf{Q}$–algebra with

$$\psi(E_\alpha) = E_\alpha, \qquad \psi(F_\alpha) = F_\alpha, \qquad \psi(K_\alpha) = K_\alpha^{-1}, \qquad \psi(q) = q^{-1}$$

for all $\alpha \in \Pi$. We have $\psi^2 = 1$. Using ψ we can state (cf. [LQG], 11.10):

Proposition: *There is for each $b \in \mathcal{B}(\infty)$ a unique element $G(b) \in \mathcal{L}_{\mathbf{Z}}(\infty)$ of the same weight as b with $\psi G(b) = G(b)$ and $G(b) + q\mathcal{L}_{\mathbf{Z}}(\infty) = b$.*

I should probably have said earlier that $\mathcal{L}(\infty)$ and $\mathcal{L}_{\mathbf{Z}}(\infty)$ are the direct sums of their weight spaces (with respect to the action of the K_α by conjugation), and that all elements in $\mathcal{B}(\infty)$ are weight vectors.

3.12. We are almost done. Set $\mathcal{B} = \{G(b) \mid b \in \mathcal{B}(\infty)\}$. Then (cf. [LQG], 11.10):

Theorem: a) *$\mathcal{B}$ is a basis of $\mathcal{L}_{\mathbf{Z}}(\infty)$ over $\mathbf{Z}[q]$, of U_A^- over A, and of U^- over k.*

b) *For each dominant $\lambda \in X$ the bv_λ with $b \in \mathcal{B}$ and $bv_\lambda \neq 0$ are a basis of $L(\lambda)$ over k.*

In order to get the basis $\mathbf{B}$ from Theorem 3.6 we regard $\mathbf{C}$ as an A–algebra via $q \mapsto 1$. One has then an isomorphism $U_A^- \otimes_A \mathbf{C} \xrightarrow{\sim} U(\mathfrak{n}^-)$. Then $\mathbf{B}$ is the set of the images of the $b \otimes 1$ with $b \in \mathcal{B}$. In order to prove Theorem 3.6 one then has to check that the A–module $L(\lambda)_A$ generated by all bv_λ with $b \in \mathcal{B}$ is a U_A–submodule of $L(\lambda)$ with $L(\lambda)_A \otimes_A \mathbf{C} \simeq V(\lambda)$ as a $U(\mathfrak{g})$–module. (See [LQG], 11.19 for more details.)

It is not easy to describe $\mathcal{B}$ explicitly. For type A_1 one gets of course all $F_\alpha^{(n)}$ with $n \geq 0$ where $\Pi = \{\alpha\}$. For type A_2 with $\Pi = \{\alpha, \beta\}$ one checks (cf. [LQG], 11.17) that $\mathcal{B}$ consists of all $F_\alpha^{(i)} F_\beta^{(l)} F_\alpha^{(j)}$ with $i + j \leq l$ and of all $F_\beta^{(i)} F_\alpha^{(l)} F_\beta^{(j)}$ with $i + j \leq l$. However, each $F_\alpha^{(i)} F_\beta^{(i+j)} F_\alpha^{(j)} = F_\beta^{(j)} F_\alpha^{(i+j)} F_\beta^{(i)}$ should be taken only once.

References

[AJS] H. H. Andersen, J. C. Jantzen, W. Soergel: Representations of quantum groups at a pth root of unity and of semisimple groups in characteristic p: independence of p, *Astérisque* **220** (1994)

[Bou] N. Bourbaki: *Groupes et algèbres de Lie, Chap. 4, 5 et 6,* Paris 1968 (Hermann)

[Ca] R. W. Carter: Introduction to algebraic groups and Lie algebras, *these proceedings,*

[CP] V. Chari, A. N. Pressley: *A Guide to Quantum Groups,* Cambridge etc. 1994 (Cambridge Univ.)

[DK1] C. De Concini, V. G. Kac: Representations of quantum groups at roots of 1, pp. 471–506 in: A. Connes et al. (eds.), *Operator Algebras,*

Unitary Representations, Enveloping Algebras, and Invariant Theory (Colloque Dixmier), Proc. Paris 1989 (Progress in Mathematics **92**), Boston etc. 1990 (Birkhäuser)

[DK2] C. De Concini, V. G. Kac: Representations of quantum groups at roots of 1: Reduction to the exceptional case, pp. 141–149 in: A. Tsuchiya, T. Eguchi, M. Jimbo (eds.), *Infinite Analysis, Part A,* Proc. Kyoto 1991 (Advanced Series in Mathematical Physics **16**), River Edge, N. J., 1992 (World Scientific)

[DKP] C. De Concini, V. G. Kac, C. Procesi: Quantum coadjoint action, *J. Amer. Math. Soc.* **5** (1992), 151–189

[Dix] J. Dixmier: *Algèbres Enveloppantes,* Paris etc. 1974 (Gauthier-Villars)

[Do] S. Donkin: Introduction to Lusztig's conjectured character formula, *these proceedings,*

[Dr] V. G. Drinfel'd: Hopf algebras and the quantum Yang-Baxter equation, *Soviet Math. Doklady* **32** (1985), 254–258, (russ. orig.:) Алгебры Хопфа и квантовое уравнение Янга-Бакстера, Докл. Акад. Наук СССР **283** (1985), 1060–1064

[Hu] J. E. Humphreys: *Introduction to Lie Algebras and Representation Theory* (Graduate Texts in Mathematics **9**), New York etc. (3rd printing) 1980 (Springer)

[LQG] J. C. Jantzen: *Lectures on Quantum Groups* (Graduate Studies in Mathematics **6**), Providence, RI, 1996 (Amer. Math. Soc.)

[Ji] M. Jimbo: A q–difference analogue of $U(\mathfrak{g})$ and the Yang-Baxter equation, *Lett. Math. Phys.* **10** (1985), 63–69

[Jo] A. Joseph: *Quantum Groups and Their Primitive Ideals* (Ergebnisse der Mathematik (3) **29**), Berlin etc. 1995 (Springer)

[Ka] C. Kassel: *Quantum Groups* (Graduate Texts in Mathematics **155**), New York etc. 1995 (Springer)

[Lu1] G. Lusztig: Quantum groups at roots of 1, *Geom. Dedicata* **35** (1990), 89–114

[Lu2] G. Lusztig: *Introduction to Quantum Groups* (Progress in Mathematics **110**), Boston etc. 1993 (Birkhäuser)

[Ro] R. Rouquier: Weyl groups, affine Weyl groups, and reflection groups, *these proceedings,*

[RŠ] A. N. Rudakov, I. R. Shafarevich: Irreducible representations of a simple three-dimensional Lie algebra over a field of finite characteristic, *Math. Notes Acad. Sci. USSR* **2** (1967), 760–767, (russ. orig.:) Неприводимые представления простой трехмерной алгебры Ли над полем конечной характеристики, Матем. заметки **2** (1967), 439–454.

Introduction to the subgroup structure of algebraic groups

Martin W. Liebeck

Imperial College, London SW7 2BZ, UK

Introduction

The purpose of this article is to discuss various results concerning the subgroups of simple algebraic groups G and of the corresponding finite groups of Lie type G^F (where F is a Frobenius morphism). There are five sections. The first contains some background on simple groups, automorphisms and reductive subgroups. In the second section we present material on two important classes of subgroups which contain a maximal torus of G: the parabolic subgroups, and the reductive "subsystem" subgroups. Section 3 contains a discussion of unipotent classes, and of subgroups of G containing various particular types of such elements. In section 4 we concentrate on closed subgroups of classical groups G. We present a recent reduction theorem which shows that any such subgroup either lies in a member of a class of naturally defined "geometric" subgroups of G, or is essentially a quasisimple group acting irreducibly on the natural module for G. Use of this result, together with a standard process involving Lang's theorem for linking finite and algebraic groups, yields a new proof of a well known reduction theorem of Aschbacher for finite classical groups, which we discuss. In the final section 5, we describe the picture for exceptional groups G. Again, there is a reduction theorem, reducing the study of subgroups H to the case where H is almost simple, and we sketch also the substantial body of recent results concerning the latter case.

1 Generalities

Let K be an algebraically closed field of characteristic p (allowing $p = 0$). The simple algebraic groups over K were classified by Chevalley [Ch], and fall into the following families:

$$\begin{array}{rl} \text{classical types}: & A_n(K), \quad B_n(K), \quad C_n(K), \quad D_n(K) \\ \text{(examples}: & SL_{n+1}(K), \quad SO_{2n+1}(K), \quad Sp_{2n}(K), \quad SO_{2n}(K)) \end{array}$$

$$\text{exceptional types}: \; G_2(K),\, F_4(K),\, E_6(K),\, E_7(K),\, E_8(K)$$

For each fixed type (e.g. $A_n(K)$ with n fixed), there may be several pairwise non-isomorphic simple algebraic groups, which can usually (but not always) be distinguished by their centres. For example, when $p = 0$ or p does not divide $n + 1$, the groups of type $A_n(K)$ are $SL_{n+1}(K)$ and its quotients by subgroups of the group of scalar matrices; the group with the largest centre, $SL_{n+1}(K)$, is the simply connected group of type $A_n(K)$, while the group with trivial centre, $PSL_{n+1}(K)$, is the adjoint group of this type. (When p divides $n + 1$, more subtle considerations are required to distinguish the different groups of type $A_n(K)$ - see for instance [Ca3, 1.11].)

Let G be a simple algebraic group over K, and fix a maximal torus T of G. A root subgroup of G is a 1-dimensional T-invariant unipotent subgroup $U_\alpha = \{U_\alpha(c) : c \in K\}$, where $\alpha \in X(T) = \mathrm{Hom}(T, K^*)$, and for $t \in T, c \in K$,

$$U_\alpha(c)^t = U_\alpha(\alpha(t)c).$$

The root subgroups U_α generate G, and the collection of roots α forms the root system Φ of G. The Weyl group $W = N_G(T)/T$ is a finite group generated by reflections corresponding to roots in Φ. Associated with the root system Φ is the Dynkin diagram of G, which has nodes labelled by a set Δ of fundamental roots $\alpha_1, \ldots, \alpha_n$, and edges determined by the inner products among these roots. If α_0 is the highest root in Φ – that is, $\alpha_0 = \sum c_i\alpha_i$ with $\sum c_i$ maximal – then the adjoining of $-\alpha_0$ to the Dynkin diagram (with appropriate edges) gives the extended Dynkin diagram of G. Full descriptions of root systems and diagrams (and many other related things) can be found in [Bou, p.250].

Next we describe the automorphism group of G as an abstract group. In order to do this, we first list various examples of automorphisms.

(1) *Inner automorphisms.*

(2) *Graph automorphisms* of types A_n, D_n, D_4, E_6. For these types, the Dynkin diagram possesses a symmetry of order 2,2,3,2 respectively, which extends linearly to a permutation ρ of the root system Φ. The graph automorphism corresponding to the symmetry ρ is given by

$$U_\alpha(c) \to U_{\rho(\alpha)}(\epsilon_\alpha c) \quad (\alpha \in \Sigma(G), c \in K),$$

where each ϵ_α is ± 1 (see [St1, p.156]).

(3) *Field automorphisms.* For each automorphism ϕ of the field K, there is a corresponding field automorphism of G given by

$$U_\alpha(c) \to U_\alpha(c^\phi) \quad (\alpha \in \Phi, c \in K).$$

Note that the field automorphisms which are morphisms are those of the form $U_\alpha(c) \to U_\alpha(c^q)$, where q is a power of p. We shall write σ_q for this particular field automorphism.

(4) *More graph automorphisms* of types $B_2\,(p=2)$, $F_4\,(p=2)$, $G_2\,(p=3)$. For these types the Dynkin diagram has a symmetry of order 2, extending to a permutation ρ of Φ which interchanges long and short roots. This time the corresponding graph automorphism of G is given by

$$U_\alpha(c) \to U_{\rho(\alpha)}(\epsilon_\alpha c^{p(\alpha)}),$$

where $p(\alpha)=1$ if α is a long root, $p(\alpha)=p$ if α is a short root, and $\epsilon_\alpha = \pm 1$ (see [St1, p.156]).

By [St1, Theorem 30], the group of automorphisms of G as abstract group is generated by the automorphisms under (1)–(4) above. On the other hand, the automorphism group of G as algebraic group (i.e. the group of automorphisms α such that α and α^{-1} are both morphisms) is generated by the automorphisms just of types (1) and (2).

The Frobenius morphisms of G are the conjugates of σ_q and of $\tau\sigma_q$, where τ is a graph automorphism and q is a power of p. By [St2, 10.13], these are precisely those surjective morphisms F of G such that the fixed point group G^F is finite. When G is of adjoint type, $O^{p'}(G^F)$ is usually simple, and is a simple group of Lie type over $\mathbb{F}_q$.

We conclude this section with a brief discussion of the closed connected subgroups of G. Let M be such a subgroup. The largest connected unipotent normal subgroup of M, denoted $R_u(M)$, is called the unipotent radical of M. The factor group $M/R_u(M)$ is a connected reductive group, hence is equal to a commuting product DZ, where D is connected semisimple and Z is a torus.

Suppose now that $Q_1 = R_u(M) \neq 1$. Define

$$Q_2 = R_u(N_G(Q_1)^0),\ Q_3 = R_u(N_G(Q_2)^0), \ldots, Q_i = R_u(N_G(Q_{i-1})^0).$$

Evidently the chain $Q_1 \leq Q_2 \leq \ldots \leq Q_i \leq \ldots$ of connected unipotent subgroups becomes stationary, at a subgroup Q_r with $Q_r = R_u(N_G(Q_r)^0)$. A basic result of Borel and Tits [BT, 2.3] says that for such a subgroup Q_r, $N_G(Q_r)$ is necessarily a parabolic subgroup of G. Thus we have

Theorem 1.1 ([BT]) *If M is a closed connected subgroup of G, then either M is reductive, or M lies in a (canonically defined) parabolic subgroup of G.*

As a consequence of this result, many questions on subgroups can be resolved by studying the classes of reductive and parabolic subgroups of G.

2 Subgroups containing a maximal torus

Continue to assume that T is a maximal torus of the simple algebraic group G. We begin this section with a discussion of the reductive subgroups containing T.

Subsystem subgroups

A *subsystem subgroup* of G is defined to be a connected semisimple subgroup which is normalized by a maximal torus. Such subgroups are determined by the root system Φ, as follows. Let Σ be a closed subsystem of Φ (that is, Σ is closed under taking negatives, and if $\alpha, \beta \in \Sigma$ and $\alpha + \beta \in \Phi$, then $\alpha + \beta \in \Sigma$). Then Σ is itself a root system, and the subgroup $G(\Sigma) = \langle U_\alpha : \alpha \in \Sigma \rangle$ is a T-invariant semisimple subgroup with root system Σ. Thus $G(\Sigma)$ (and all its conjugates) is a subsystem subgroup. Indeed, except for some special cases which occur only when $p = 2$ or 3 and there is more than one root length, these are the only subsystem subgroups. Moreover, any connected reductive subgroup of G containing T is of the form DZ, where D is a subsystem subgroup and Z a torus commuting with D.

There is an elegant algorithm due to Borel and de Siebenthal [BS] which determines all closed subsystems of Φ, and proceeds as follows. Start with the Dynkin diagram of G, and adjoin $-\alpha_0$ to form the extended diagram. Remove any collection of nodes from this, and repeat the process with all connected components of the resulting graph. The diagrams obtained in this way are the Dynkin diagrams of subsystems, and there is one conjugacy class of subsystem subgroups for each such subsystem. In particular, G possesses only finitely many conjugacy classes of subsystem subgroups. Lists of subsystems for G of exceptional type can be found in [Ca1].

For example, by removing appropriate nodes from the extended E_8 diagram (see [Bou, p. 250]), we can obtain subsystem subgroups of E_8 of type $A_1E_7, D_8, A_8, A_2E_6, A_4A_4$ and so on.

Important examples of reductive subgroups containing maximal tori are the centralizers of semisimple elements in G. Indeed, if t is a semisimple element of G, then t is conjugate to an element of T, so $C_G(t)^0$ contains a conjugate of T; thus $C_G(t)^0 = DZ$ for some subsystem subgroup D and commuting torus Z. Which subsystem subgroups occur in this way ? Here is the answer in the important special case where t has prime order.

Theorem 2.1 ([GL, 14.1]) *Let $x \in G$ be a semisimple element of prime order r, and let C' be the subsystem subgroup $(C_G(x)^0)'$. If Σ is the Dynkin diagram of the root system of C', then one of the following holds:*

(i) *Σ is obtained by deleting nodes from the Dynkin diagram of G;*

(ii) *Σ is obtained from the extended Dynkin diagram of G by deleting one node α_i, where $r = c_i$, the coefficient of α_i in the highest root α_0.*

For example, when $G = E_8$ we have

$$\alpha_0 = \begin{array}{ccccccc} 2 & 4 & 6 & 5 & 4 & 3 & 2 \\ & & 3 & & & & \end{array}$$

where the node α_i of the E_8 Dynkin diagram is labelled by its coefficient c_i. Referring to the extended Dynkin diagram [Bou, p.250], from 2.1(ii) we see that E_8 possesses the following classes of semisimple elements:

for $p \neq 2$, elements of order 2 with centralizers of type D_8 and A_1E_7;
for $p \neq 3$, elements of order 3 with centralizers of type A_8 and A_2E_6;
for $p \neq 5$, elements of order 5 with centralizer of type A_4A_4.

The primes dividing the coefficients c_i of the highest root α_0 are called *bad* primes for G. The bad primes are as follows (see [SS, I, §4]):

$$\begin{aligned} &G = A_n : \text{ none} \\ &G = B_n, C_n, D_n : \text{ only } 2 \\ &G = G_2, F_4, E_6, E_7 : 2 \text{ and } 3 \\ &G = E_8 : 2, 3 \text{ and } 5. \end{aligned}$$

We now discuss the subsystem subgroups of the finite groups G^F, where F is a Frobenius morphism. Let $D = G(\Sigma)$ be a subsystem subgroup of G with root system Σ. By a subsystem subgroup of G^F we mean a subgroup of the form $(D^g)^F$, where D^g is an F-stable G-conjugate of D. An application of Lang's theorem shows that the G^F-classes of such subgroups are in 1-1 correspondence with the F-classes in $N_G(D)/N_G(D)^0$, which is isomorphic to $N_W(\Sigma)/W(\Sigma)W(\Sigma^\perp)$, the group of symmetries induced by W on the Dynkin diagram of Σ (see [Ca2, §2]).

As an example, let $G = SL_6$ (type A_5), and let $D = (SL_2)^3$, a subsystem subgroup of type $(A_1)^3$. Here $N_G(D)/N_G(D)^0 \cong S_3$, permuting the factors SL_2. Take $F = \sigma_q$, so $G^F = SL_6(q)$. As F acts trivially on W, the F-classes in S_3 are just the conjugacy classes, with representatives 1, (12) and (123). If D^g is an F-stable conjugate of D, then $g^F g^{-1} \in N_G(D)$, hence maps to a conjugate w of one of the above representatives. The action of F on D^g is like that of wF on D; thus the three G^F-classes of subsystem subgroups of G^F arising from conjugates of D are as follows:

(1) $w = 1 \ : \ D^F = SL_2(q)^3$;

(2) $w = (12) \ : \ F$-action on D^g is $(g_1, g_2, g_3) \to (g_2^{(q)}, g_1^{(q)}, g_3^{(q)})$ for $g_i \in SL_2$, so

$$(D^g)^F = SL_2(q^2) \times SL_2(q);$$

(3) $w = (123) \ : \ F$-action is $(g_1, g_2, g_3) \to (g_3^{(q)}, g_1^{(q)}, g_2^{(q)})$, and

$$(D^g)^F = SL_2(q^3).$$

Parabolic subgroups

From some points of view (such as conjugacy classes, overgroups, intersections and so on), the parabolic subgroups of G are well understood. However, there are some interesting aspects of their internal structure which are perhaps less well known. We sketch one of these aspects.

Let Δ denote a system of fundamental roots in the root system Φ, and let $J \subseteq \Delta$. Write Φ_J for the root system $\Phi \cap \mathbb{Z}J$, and define the parabolic subgroup

$$P_J = \langle T, U_\alpha, U_{-\beta} : \alpha \in \Phi^+, \beta \in \Phi_J^+ \rangle.$$

Then $P_J = QL$, where

$$Q = \langle U_\alpha : \alpha \in \Phi^+ \backslash \Phi_J^+ \rangle, \quad L = \langle T, U_\beta : \beta \in \Phi_J \rangle.$$

The group Q is the unipotent radical of P_J, and L is a Levi subgroup. The structure of L is transparent - it is a commuting product of a torus and a semisimple group having Dynkin diagram J. The action of L on Q is much less transparent. Take an L-composition series

$$1 = Q_0 \lhd Q_1 \lhd \ldots \lhd Q_r = Q.$$

Results in [ABS] establish that each factor Q_i/Q_{i-1} has the structure of an irreducible KL-module, of which the high weight can be calculated within the root system Φ. Moreover, by a theorem of Richardson [Ri], L has finitely many orbits on the vectors of each of these irreducible modules (and in particular, has an open dense orbit). Irreducible modules for algebraic groups with only finitely many orbits have been classified in [Kac, GLMS]; these modules are rather rare, and quite a large proportion of them occur within parabolics, as above.

It may be helpful to give an example. Let $G = E_8$, and let P_J be the parabolic subgroup with Levi subgroup A_7T_1. Then the above composition series is $1 < Q_1 < Q_2 < Q$, and the irreducible quotient modules are $V_8, \Lambda^2 V_8$ and $\Lambda^3 V_8$, where V_8 is the natural 8-dimensional module for L. Consequently, GL_8 has finitely many orbits on V_8 and $\Lambda^2 V_8$ (which is obvious), and also on $\Lambda^3 V_8$ (which is far from obvious).

Representatives (or even just the numbers) of orbits for the modules occuring as above are not available in any systematic form. However, there is one special case in which there is an elegant description [RRS] - that in which Q is abelian (in which case Q is itself an irreducible module for L). A rather surprising phenomenon occurs in this case:

Theorem 2.2 ([RRS, 2.10]) *If $P = QL$ is a parabolic subgroup such that $Q = R_u(P)$ is abelian, then the number of orbits of L on Q is equal to the number of $P\backslash G/P$-double cosets.*

Indeed, there is a natural correspondence between the orbits and the double cosets in this case.

The matters discussed above have been taken much further by Röhrle and others (see [Rö] for example).

3 Unipotent elements

Recall the Jordan decomposition for elements of the simple algebraic group G: each $x \in G$ has a unique expression $x = x_s x_u$, where x_s is semisimple, x_u is unipotent, and x_s, x_u commute. We have already commented briefly on the classes and centralizers of semisimple elements. In this section we turn to unipotent elements, and discuss some results concerning their classes and overgroups. We begin with a theorem of Lusztig.

Theorem 3.1 ([Lu]) *The number of conjugacy classes of unipotent elements in a simple algebraic group is finite.*

For example, in $SL_n(K)$, the unipotent classes are parametrized by Jordan canonical forms, so the number of classes is the number of partitions of n.

There are complete descriptions of the conjugacy classes of unipotent elements in all types of simple algebraic group; see [Ca3, Chapter 5] for discussion and references. Here we highlight two "extreme" classes:

(1) *long root* elements: these are non-identity elements of root subgroups U_α corresponding to a long root $\alpha \in \Phi$. (For example in SL_n or Sp_n, these are the transvections.) For a long root element u, we have $C_G(u) = P'$, where P is a parabolic subgroup (usually maximal).

(2) *regular* unipotent elements: these are conjugates of $\prod_{\alpha\in\Delta} U_\alpha(1)$, where Δ is a fundamental system in Φ. (In SL_n, they are the unipotent elements with a single Jordan block.) For a regular unipotent element u we have $C_G(u) = U \times Z(G)$, where U is an abelian unipotent group of dimension equal to the rank of G.

The class of regular unipotent elements is dense in the set of all unipotent elements of G. In a (rather weak) sense, an arbitrary unipotent class in G lies somewhere "in between" the above two classes. To explain this, first extend the definition of regular by defining a unipotent element u to be *semiregular* if $C_G(u) = Q \times Z(G)$ with Q a unipotent subgroup. If u is any unipotent element of G, then there is a connected semisimple subsystem subgroup $D = D_1 \dots D_r$ of G (all D_i simple) such that $u = u_1 \dots u_r$, with each u_i a semiregular element of D_i.

There is quite a large body of literature concerning the subgroups of G and G^F containing an element of some special type. Probably the most studied is the case where the element is a long root element. For the finite groups, complete lists of subgroups containing long root elements have been obtained by Kantor [Ka] for classical groups, and by Cooperstein [Co1, Co2, Co3] for exceptional groups. A more conceptual approach was found in [LS2]; this contains the following result for simple algebraic groups G, from which a corresponding result for the finite groups G^F is deduced.

Theorem 3.2 ([LS2]) *Let X be a simple closed connected subgroup of G such that $|U \cap X| > 2$ for some long root subgroup U of G. Then there is a simple subsystem subgroup Y of G such that $X \leq Y$ and one of the following holds:*
(i) $X = Y$;
(ii) $X = C_Y(\tau)$ *for some graph automorphism* τ *of* Y;
(iii) $(X, Y) = (G_2, B_3)$ *or* (C_4, E_6) *(with* $p = 2$ *in the latter case).*

Note that [LS2] also contains a result covering the case where $|U \cap X| = 2$.

The connection with subsystem subgroups makes it easy to study conjugacy classes of, centralizers of, and restrictions of representations to, reductive subgroups X generated by long root elements.

We now move on to regular unipotent elements. Of course, every such element lies in a parabolic subgroup, but one would expect there to be very few reductive subgroups containing a regular unipotent element. In $SL_n(K)$, provided $p = 0$ or $p \geq n$, there is such a reductive subgroup: the image of $SL_2(K)$ in an irreducible representation over K of high weight $n - 1$. And indeed, every simple algebraic group G contains a "regular" A_1 subgroup in which each non-identity unipotent element is regular in G, provided $p = 0$ or $p \geq h$, the Coxeter number of G. The next result shows that, at least among the maximal connected subgroups of G, these are usually the only reductive overgroups of regular unipotents.

Theorem 3.3 ([SaSe]) *Suppose X is a maximal connected subgroup of G and X contains a regular unipotent element of G. Then one of the following holds:*
(i) *X is a parabolic subgroup;*
(ii) *X is a regular A_1 subgroup ($p = 0$ or $p \geq h$);*
(iii) $X = C_G(\tau)$ *with* τ *a graph automorphism, or* $(X, G) = (G_2, B_3)$.

For arbitrary unipotent elements, the following theorem of Testerman is the most general result available.

Theorem 3.4 ([Te]) *Assume $p > 0$ and p is a good prime for G. Then each element of order p in G lies in a closed connected subgroup of type A_1.*

4 Classical groups

In this section we shall describe reduction theorems concerning the subgroups of algebraic and finite classical groups. Let V be a vector space of finite dimension n over the algebraically closed field K, and write

$$G = Cl(V)$$

to indicate that G is one of the classical groups $SL(V), Sp(V)$ or $SO(V)$. The main results take the following form. We define a collection $\mathcal{C}$ of "natural" subgroups of G (or G^F). Then any closed subgroup of G (or of G^F) either lies in a member of $\mathcal{C}$, or is essentially a quasisimple group acting irreducibly on V. We call such a result a "reduction theorem", since for many purposes it reduces the study of subgroups to the study of irreducible representations of quasisimple groups. The reduction theorem for the finite groups G^F is deduced from the reduction theorem for G. This gives a new proof of a well known result of Aschbacher [As] on subgroups of finite classical groups.

We begin with an elementary reduction theorem for connected subgroups of G.

Proposition 4.1 *Let H be a closed connected subgroup of G. Then one of the following holds:*

(i) *$H \leq G_U$, the stabilizer in G of a proper nonzero subspace U of V;*

(ii) *$V = V_1 \otimes V_2$ and H lies in a subgroup of G of the form $Cl(V_1) \circ Cl(V_2)$ acting naturally on the tensor product;*

(iii) *H is simple, and acts irreducibly and tensor-indecomposably on V.*

Remarks 1. When $G = Sp(V)$ or $SO(V)$, the subspace U in (i) can be taken to be totally singular or non-degenerate, or, in the case where $(G, p) = (SO(V), 2)$, nonsingular of dimension 1. Also, the precise tensor product subgroups occurring in (ii) are of the following types:

$$SL \otimes SL < SL, \quad Sp \otimes SO < Sp, \quad Sp \otimes Sp < SO, \quad SO \otimes SO < SO.$$

2. The subgroups H in (iii) are usually maximal in one of the groups $Cl(V)$, with an explicit list of exceptions (see [Se1, Theorem 1]).

Here is a sketch proof of 4.1. If H is non-reductive, then the Borel-Tits theorem 1.1 implies that $H \leq P$ for some parabolic subgroup P; and $P \leq G_U$ for some (totally singular) subspace U, giving conclusion (i). Thus we may assume that H is reductive; we may also take it that H is irreducible on V (otherwise (i) holds again). Then $Z(H)$ acts as scalars on V, hence is finite. It follows that H is semisimple, say $H = H_1 \ldots H_t$, a commuting product of simple groups H_i. If $t > 1$ then H preserves a tensor decomposition $V = V_1 \otimes \ldots \otimes V_t$, and it is not hard to see that (ii) holds. And if $t = 1$ then H is simple and we obtain (iii).

Proposition 4.1 only covers connected subgroups, and hence is not applicable in some important situations - in particular it will not apply to the study of finite subgroups. What is required is the generalization given in Theorem 4.2 below, a reduction theorem for arbitrary closed subgroups of G.

In order to state the result, we need to define a number of classes of subgroups M of the classical group $G = Cl(V)$:

Class $\mathcal{C}_1$. *Subspace stabilizers* $M = G_U$, where U is a proper nonzero subspace of V; moreover, U is totally singular or non-degenerate, or, in the case where $(G, p) = (SO(V), 2)$, nonsingular of dimension 1.

Class $\mathcal{C}_1'$. *Subspace-pair stabilizers:* here $G = SL(V)$ and $M = G_{U,W}$, where U, W are proper subspaces of V such that $\dim U + \dim W = \dim V$ and either $U \subseteq W$ or $U \cap W = 0$; also, $\dim U \neq \dim W$.

Class $\mathcal{C}_2$. *Stabilizers of orthogonal decompositions*: here $V = \bigoplus_1^t V_i$, where $t > 1$ and the subspaces V_i are mutually orthogonal and isometric, and $M = G_{\{V_1,\dots,V_t\}}$; in general, $M \cong Cl(V_1)$ wr Sym_t (where Sym_t denotes the symmetric group of degree t).

Class $\mathcal{C}_3$. *Stabilizers of totally singular decompositions*: here $G = Sp(V)$ or $SO(V)$, $V = W \oplus W'$ where W, W' are maximal totally singular subspaces, and $M = G_{\{W,W'\}}$; if $\dim V = 2m$, then $M \cong GL_m.2$ or GL_m (the latter only when $G = SO(V)$ and m is odd).

Class $\mathcal{C}_4$. *Tensor product subgroups*: either

(i) $V = V_1 \otimes V_2$ and $M = Cl(V_1) \circ Cl(V_2)$ as in 4.1(ii), or

(ii) $V = \bigotimes_1^t V_i$ with $t > 1$, the V_i mutually isometric, and we have $M = N_G(\prod Cl(V_i))$, where $\prod Cl(V_i)$ acts naturally on the tensor product; in general, we have $M = (\prod Cl(V_i)).\mathrm{Sym}_t$.

Class $\mathcal{C}_5$. *Finite local subgroups*: let r be a prime different from p, and let R be an extraspecial r-group of order r^{1+2m}, or, when $r = 2$, a central product of such a group with a cyclic group of order 4. Every faithful irreducible representation of R has degree r^m. Take $n = \dim V$ to be r^m, and embed R as a subgroup of $G = Cl(V)$ via one of these representations, where $Cl(V)$ is $SL(V)$ if r is odd or $R = C_4 \circ 2^{1+2m}$, and $Cl(V)$ is $Sp(V)$ or $SO(V)$ otherwise. Then the members of the class $\mathcal{C}_5$ are the subgroups $N_G(R)$ with R as above. If $G_1 = N_{SL(V)}(G)$, $Z = Z(G)$ and $M = N_{G_1}(R)$, then $\bar{M} = M/Z \cong C_{\mathrm{Aut}(R)}(Z(R))$. Thus the subgroups in the class are as follows:

$$\begin{aligned} G &= SL_{r^m}, & \bar{M} &= r^{2m}.Sp_{2m}(r) \\ G &= Sp_{2^m}, & \bar{M} &= 2^{2m}.O_{2m}^-(2) \\ G &= SO_{2^m}, & \bar{M} &= 2^{2m}.O_{2m}^+(2). \end{aligned}$$

Definition We write $\mathcal{C}(G)$ for the collection $\mathcal{C}_1 \cup \mathcal{C}_1' \cup \ldots \cup \mathcal{C}_5$ of subgroups of G defined above.

We can now state the reduction theorem for G. If $p > 0$, let F be a Frobenius morphism of G such that G^F is a finite classical group.

Theorem 4.2 ([LS5]) *Let $G = Cl(V)$ be a classical algebraic group, and let H be a closed subgroup of G. Then one of the following holds:*

(i) *H is contained in a member of $\mathcal{C}(G)$, which is F-stable if H is F-stable;*

(ii) *modulo scalars, H is almost simple, and $E(H)$ (the unique quasisimple normal subgroup of H) acts irreducibly on V.*

We remark that in the theorem, H may of course be finite or infinite; if it is infinite, the conclusion of (ii) can be strengthened to say that $E(H)$ also acts tensor-indecomposably on V.

The proof of 4.2 is naturally somewhat more complicated than that of 4.1, but uses only elementary linear algebra, together with a small amount of Lie algebra theory.

We now outline the use of Theorem 4.2 in proving a corresponding reduction theorem for finite classical groups; details can be found in [LS5]. Continue to assume that $G = Cl(V) = SL(V), Sp(V)$ or $SO(V)$, and let F be a Frobenius morphism of G such that $G^F = SL_n(q)$, $SU_n(q)$, $Sp_n(q)$ or $SO_n^{\pm}(q)$; write $G^F = Cl_n(q)$. The natural (finite) module for G^F is $V^{F^u} = V_n(q^u)$, where $u = 2$ if G^F is unitary and $u = 1$ otherwise.

Let H be a subgroup of G^F. Then 4.2(i) or (ii) holds for H. If 4.2(i) holds then $H \leq M^F$ for some $M \in \mathcal{C}(G)$; thus to obtain a reduction theorem for G^F, we need to analyse the G^F-classes of subgroups M^F. This can be done using Lang's theorem and some linear algebra. We sketch some of the arguments - for details, see [LS5].

If M lies in $\mathcal{C}_1$, then $M = G_U$ for some subspace U of V and we easily see that H lies in a member of the following class of subgroups of G^F:

Class $\mathcal{C}_1^F$. Stabilizers in G_F of subspaces of $V_n(q^u)$.

Now let $M \in \mathcal{C}_2$, so $M = G_{\{V_1,\dots,V_t\}}$ where $V = \bigoplus V_i$, and F permutes the factors $Cl(V_i)$. Moreover, the group of permutations induced by M^F on the factors is a subgroup of $C_{\mathrm{Sym}_t}(F)$. If $\langle F\rangle$ induces a nontrivial imprimitive subgroup of Sym_t, then M^F permutes the sums of subspaces corresponding to each block of imprimitivity. Hence, replacing M by another member of $\mathcal{C}_2$ if necessary, we may assume that either $\langle F\rangle$ induces a trivial group, or t is prime and $\langle F\rangle$ induces a transitive group. In the first case, in general $M^F \cong Cl_m(q)$ wr Sym_t (where $n = mt$); in the second, roughly speaking, $M^F \cong Cl_m(q^t).t$. Thus we obtain the following classes of subgroups of G^F:

Class $\mathcal{C}_2^F$. Stabilizers in G^F of orthogonal decompositions of $V_n(q^u)$ (type $Cl_m(q)$ wr Sym_t, where $n = mt$).

Class $\mathcal{C}_2'^F$. Extension field subgroups (type $Cl_m(q^t).t$ with $n = mt$ and t prime).

Next, suppose that $M \in \mathcal{C}_3$, so M is the stabilizer of a totally singular decomposition $V = W \oplus W'$. Here $M = GL_{\frac{n}{2}}.2$ (or $GL_{\frac{n}{2}}$), and M^F is $GL_{\frac{n}{2}}(q).2$ or $GU_{\frac{n}{2}}(q).2$ (or possibly a subgroup of small index in one of these); thus H lies in a member of the following class.

Class $\mathcal{C}_3^F$. Subgroups of type $GL_{\frac{n}{2}}(q).2$ and $GU_{\frac{n}{2}}(q).2$, where G^F is symplectic or orthogonal.

Now consider $M \in \mathcal{C}_4$, a tensor product subgroup. If $M = N_G(\prod Cl(V_i))$, where $V = \bigotimes V_i$ and the V_i are isometric, we argue as in the case where $M \in \mathcal{C}_2$ that either F acts trivially on the factors, or t is prime and $\langle F \rangle$ acts transitively. Thus H lies in one of the following subgroups.

Class $\mathcal{C}_4^F$. Tensor product subgroups of type $Cl_{n_1}(q) \circ Cl_{n_2}(q)$ $(n = n_1 n_2)$, or of type $N_{G^F}(\prod_1^t Cl_m(q))$ $(n = m^t)$.

Class $\mathcal{C}_4'^F$. Subgroups of type $Cl_m(q^t).t$ with $n = m^t$ and t prime.

Finally, suppose that $M \in \mathcal{C}_5$, so M is a finite local subgroup $N_G(R)$. In this case we argue that if $M^F < M$ then M lies in a member of $\mathcal{C}_1 \cup \ldots \cup \mathcal{C}_4$. Hence we may take it that $M^F = M$, and define the class

Class $\mathcal{C}_5^F$. Local subgroups $N_{G^F}(R)$, as in $\mathcal{C}_5$.

We have now outlined the proof of the reduction theorem for finite classical groups:

Theorem 4.3 ([As]) *If G^F is a finite classical group as above, and H is a subgroup of G^F, then either*

(i) *H lies in a member of $\mathcal{C}_1^F \cup \ldots \cup \mathcal{C}_5^F$, or*

(ii) *modulo scalars, H is almost simple, and $E(H)$ is absolutely irreducible on the natural module for G^F.*

Remarks 1. In fact, Aschbacher's result [As] is more detailed than that stated in 4.3. His result involves two further classes of subgroups: $\mathcal{C}_6^F$, a class of subgroups whose representations on $V_n(q^u)$ are realised over a proper subfield of F_{q^u}; and $\mathcal{C}_7^F$, the class of classical groups on $V_n(q^u)$ which lie in G^F. His conclusion in (ii) is correspondingly more refined - for example, the representation of $E(H)$ can be taken not to be realised over a proper subfield. In [LS5] similar methods to those outlined above are used to obtain the more refined result.

2. The subgroups in $\mathcal{C}_4'^F$ do not occur in [As], and indeed satisfy both conclusions (i) and (ii) of 4.3. These subgroups play a major role in the extension of Aschbacher's theorem proved in [Se2, Corollary 6].

5 Exceptional groups

Throughout this section, let G be a simple algebraic group of exceptional type G_2, F_4, E_6, E_7 or E_8 over the algebraically closed field K. We begin by discussing some results concerning connected reductive subgroups of G, for which there is now quite a satisfactory theory, at least under some mild assumptions on the characteristic p of K. We then move on to the study of arbitrary closed subgroups, which divides naturally into the cases of finite and infinite subgroups.

Connected subgroups

We begin with the following result of Seitz [Se3] on maximal connected subgroups, which underlies the entire body of work presented in this section.

Theorem 5.1 ([Se3]) *Let M be a maximal closed connected subgroup of G. If M is simple, assume that either $p = 0$ or $p > 7$. Then either*

(i) *M is a parabolic subgroup or a subsystem subgroup; or*

(ii) *G and M are as follows (one* Aut(G)*-class for each M listed, except when otherwise stated):*

$G = G_2$: $M = A_1$
$G = F_4$: $M = A_1$, G_2 *or* A_1G_2
$G = E_6$: $M = A_2$, G_2, F_4, C_4 *or* A_2G_2
$G = E_7$: $M = A_1$ *(two classes)*, A_2, A_1A_1, A_1G_2, A_1F_4 *or* G_2C_3
$G = E_8$: $M = A_1$ *(three classes)*, B_2, A_1A_2 *or* G_2F_4.

Remark The assumptions on p in [Se3, Theorem 1] are in fact much weaker than the assumption $p = 0$ or $p > 7$ in 5.1. Roughly speaking they are as follows:

if $M = A_1$, assume $p = 0$ or $p > 7$;

if M has rank 2, assume $p = 0$ or $p > 5$;

if M has rank 3 or $(G, M) = (E_8, B_4)$, assume $p \neq 2$.

In all other cases, no assumption on p is made. This remark applies to all the other results in this section in which characteristic assumptions are made.

We now turn to arbitrary connected reductive subgroups. We begin with a trivial observation on subgroups of $GL_n(K) = GL(V)$: a subgroup H of $GL(V)$ is completely reducible on V if and only if, whenever H lies in a parabolic subgroup QL of $GL(V)$ (with unipotent radical Q, Levi subgroup L), then H lies in a Q-conjugate of L. If $p = 0$, or $p \geq \dim V$, then all connected reductive subgroups of $GL(V)$ are completely reducible (see [Ja]).

Is there a similar result for subgroups of exceptional groups ? The answer is yes:

Theorem 5.2 ([LS3, Theorem 1]) *Let H be a connected reductive subgroups of the exceptional algebraic group G, and assume that $p = 0$ or $p > 7$. If H lies in a parabolic subgroup QL of G, then H is contained in a Q-conjugate of L.*

In other words, excluding some small characteristics, connected reductive subgroups are "completely reducible".

As a consequence of this result, all connected reductive subgroups of G can be determined. To state the result, we need a definition.

Definition Let $Y = Y_1 \dots Y_t$ be a semisimple connected algebraic group, with each Y_i simple. We say that a closed connected semisimple subgroup X of Y is *essentially embedded* in Y if the following hold for all i:

(i) if Y_i is a classical group with natural module V_i, then either the projection of X in Y_i is irreducible on V_i, or $Y_i = D_n$ and the projection lies in a natural subgroup $B_r B_{n-r-1}$ of Y_i, irreducible in each factor with inequivalent representations;

(ii) if Y_i is of exceptional type, then the projection of X in Y_i is either Y_i or a maximal connected subgroup of Y_i not containing a maximal torus (hence given in 5.1(ii) for $p = 0$ or $p > 7$).

Theorem 5.3 ([LS3, Theorem 5]) *Let X be a closed connected semisimple subgroup of the exceptional group G, and suppose that $p = 0$ or $p > 7$. Assume that X has no factor of type A_1. Then there is a subsystem subgroup Y of G such that X is essentially embedded in Y.*

When X has a factor A_1, there is a similar result which is somewhat more complicated to state; we refer the reader to [LS3, Theorem 7]. As a consequence, when $p = 0$ there are only finitely many conjugacy classes of closed connected semisimple subgroups in G, whereas there are infinitely many when $p > 0$.

The classical factors of subsystem groups are of small rank (at most 8), so their essentially embedded closed connected semisimple subgroups can be determined, using some representation theory. Using 5.3, the conjugacy classes and centralizers in G of all simple connected subgroups of rank at least 2 are explicitly listed in [LS3, §8]; the same is done for subgroups of type A_1 in [LT].

Using 5.3, one can work out centralizers of reductive subgroups of G, and restrictions of the Lie algebra $L(G)$ to reductive subgroups. Several interesting consequences emerge, such as the following two results.

Corollary 5.4 ([LS3, Theorem 2]) *Let X be a closed connected reductive subgroup of G, and assume that $p = 0$ or $p > 7$. Then $C_G(X)^0$ is reductive.*

Corollary 5.5 ([LS3 Theorem 3; LT]) *Suppose that X is a closed connected simple subgroup of G, and $p = 0$ or $p > 7$. Then $C_{L(G)}(X) = L(C_G(X))$.*

Closed subgroups: the infinite case

As in §4, for applications one needs to strengthen 5.1 in two ways: firstly, by dropping the assumption of connectedness on M; and secondly, by keeping track of F-stability, where F is a Frobenius morphism of G. This is achieved in the next result, which determines the maximal closed subgroups M of G of positive dimension. Surprisingly, it turns out that when M^0 does not contain a maximal torus of G, there are just three classes for which M is maximal closed but M^0 is not maximal connected:

Theorem 5.6 ([LS1, Theorem 1]) *Suppose that M is a maximal closed (F-stable) subgroup of G with $M^0 \neq 1$. If M^0 is simple, assume that $p = 0$ or $p > 7$. Then one of the following holds:*

(i) *M^0 contains a maximal torus of G;*

(ii) *M^0 is as in (ii) of 5.1;*

(iii) *$G = E_7$ and $M = (2^2 \times D_4).\mathrm{Sym}_3$ (where $M^0 = D_4 < A_7$ subsystem);*

(iv) *$G = E_8$ and $M = A_1 \times \mathrm{Sym}_5$ (where $M^0 = A_1 < A_4A_4$ subsystem);*

(v) *$G = E_8$ and $M = (A_1G_2G_2).2$ (where $M^0 = A_1G_2G_2 < F_4G_2 < G$).*

As will be seen below, this result is fundamental also to the study of finite subgroups (both of G and of G^F), and it would be very desirable to remove the assumptions on the characteristic p. Work on this is in progress, by Seitz and the author.

Closed subgroups: the finite case

Let H be a finite subgroup of the exceptional algebraic group G. We begin by describing a reduction theorem for such subgroups.

Assume that H normalizes no connected subgroup D such that $1 < D < G$. (Otherwise HD (hence H) lies in a maximal closed subgroup of positive dimension, given by 5.6 for $p = 0$ or $p > 7$.)

Consider first the case where H is local, so that $H \leq N_G(E)$ for some elementary abelian r-subgroup E of G (where r is prime). We may assume $r \neq p$, by [BT]. Some rather interesting local subgroups emerge in this situation:

Theorem 5.7 ([Bor, CLSS]) *Under the above hypotheses, one of the following holds:*

$$\begin{aligned} G = G_2: &\quad E = 2^3,\ H = N_G(E) = 2^3.SL_3(2) \\ G = F_4 \text{ or } E_6: &\quad E = 3^3,\ H = N_G(E) = 3^3.SL_3(3) \text{ or } 3^{3+3}.SL_3(3) \\ G = E_8: &\quad E = 2^5 \text{ or } 5^3,\ H = N_G(E) = 2^{5+10}.SL_5(2) \text{ or } 5^3.SL_3(5). \end{aligned}$$

Definition We call the subgroups $N_G(E)$ in the conclusion of 5.7 *exotic local* subgroups.

Now assume that our finite subgroup H of G is non-local. Thus $F^*(H) = F_1 \times \ldots \times F_t$, where the F_i are non-abelian simple groups. When $t = 1$, H is almost simple, which is our desired conclusion. So assume that $t > 1$.

At this point we apply a double centralizer argument due to Borovik. For each i, let $CC(F_i)$ be the double centralizer $C_G(C_G(F_i))$ of F_i in G. Then $\prod CC(F_i)$ is a commuting product of subgroups of G which is proper, nontrivial, and normalized by H. Therefore by assumption, $CC(F_i)$ is finite for all i. If we choose $x \in F_i$ of prime order at least 5, then $CC(F_i)$ contains $Z(C_G(x))$. By the theory of centralizers in G (see 2.1 for the case where x is semisimple), $Z(C_G(x))$ can only be finite if the order of x is a bad prime for G. Hence $G = E_8$ and each F_i is a simple $\{2,3,5\}$-group. The only such simple groups are Alt_5, Alt_6 and $U_4(2)$ (where Alt_k denotes the alternating group of degree k). Further argument now forces

$$G = E_8, \ F^*(H) = \mathrm{Alt}_5 \times \mathrm{Alt}_6.$$

Somewhat amazingly, there *is* such a subgroup H in $G = E_8$ which normalizes no proper connected subgroup, as was first shown by Borovik [Bor].

We have now sketched the proof of the reduction theorem for finite subgroups of G:

Theorem 5.8 ([Bor, LS1]) *Let H be a finite subgroup of the exceptional algebraic group G. Then one of the following holds:*

(i) *H is almost simple;*

(ii) *H is contained in a proper maximal closed subgroup of positive dimension in G (given by 5.6 if $p = 0$ or $p > 7$);*

(iii) *H is contained in an exotic local subgroup;*

(iv) *$G = E_8$ and $H = (\mathrm{Alt}_5 \times \mathrm{Alt}_6).2^2$.*

Remark A more complete reduction theorem for maximal subgroups of the finite groups G^F (F a Frobenius morphism), again with no characteristic assumptions, can be found in [LS1, Theorem 2].

In view of the above reduction theorem, attention now focuses on the finite (almost) simple subgroups of G. There is a considerable literature on this, which we now proceed to discuss.

Finite simple subgroups

We begin by discussing the case where $K = \mathbb{C}$ (so G is a complex exceptional Lie group).

Let S be a finite simple subgroup of $G = G(\mathbb{C})$. As G has a nontrivial complex representation of degree at most 248, so does S. However, [LaSe] gives strong lower bounds for the degrees of nontrivial complex representations of finite groups of Lie type (and bounds are also available for alternating and sporadic groups). For example, if $S = L_n(q)$ and χ is a nontrivial complex character of S, then $\chi(1) \geq q^{n-1} - 1$, and hence $q^{n-1} - 1 \leq 248$; this gives a small list of possibilities for (n, q) with $L_n(q) < G$. In this way, one quickly reduces to a finite list of possibilities for S.

Cohen, Griess, Serre, Wales and others have taken this analysis much further, and there are now complete lists of isomorphism types of simple subgroups S of G, with just a few open cases remaining (see [CW1, CW2, CG, Ser]). However, the conjugacy classes of these subgroups remain largely undetermined. Could there even be infinitely many G-classes of simple subgroups of a given isomorphism type? The answer to this question, at least, is no, as the next result shows; this goes back to Weil [We], and is also proved in [Sl].

Theorem 5.9 *Let H be a finite group, and X a linear algebraic group over an algebraically closed field of characteristic p, where either $p = 0$ or p does not divide $|H|$. Then, up to X-equivalence, there are only finitely many homomorphisms from H into X.*

We mention one well known conjecture relating to the problem of this section. This is Kostant's conjecture, that if h is the Coxeter number of the exceptional algebraic group $G = G(\mathbb{C})$, then G contains a copy of $PSL_2(2h+1)$. (Note that $h = 6, 12, 12, 18, 30$ according as $G = G_2, F_4, E_6, E_7, E_8$, respectively, so $2h + 1$ is a prime power.) This conjecture was verified case-by-case [CGL, KR] using a computer; and an elegant computer-free proof has been given by Serre [Ser].

We conclude with a discussion of finite simple subgroups S of exceptional algebraic groups $G = G(K)$ when the algebraically closed field K has positive charateristic p. The analysis breaks up into two cases:

(1) the *generic* case, in which $S = S(p^e)$ is of Lie type in characteristic p, and

(2) the *non-generic* case, in which S is alternating, sporadic or of Lie type in p'-characteristic.

For the non-generic case, similar methods to those used for the complex case $K = \mathbb{C}$ are available, and work on this is in progress by Seitz and the author.

Finally we discuss the generic case. Here the aim is to lift the embedding $S(p^e) < G$ to an embedding $\bar{S} < G$, where $\bar{S}$ is a connected simple group of the same type as S. Once such a lifting is achieved, we can then use results such as 5.1, 5.3 and 5.6 to identify the connected subgroup $\bar{S}$, and hence also S.

The first such lifting result was proved by Seitz and Testerman [ST], under the assumption that the characteristic p is large. More recently, Seitz and the author [LS4] have obtained a lifting result, under assumptions only on the field size p^e:

Theorem 5.10 ([LS4]) *Let $S = S(p^e) < G$, and assume*

$$\begin{array}{ll} p^e > 9 & \text{if } S \neq L_2(p^e), {}^2B_2(p^e), {}^2G_2(p^e), \\ p^e > t(G) & \text{otherwise} \end{array}$$

(where $t(G)$ is a constant defined in terms of the root system of G). Then S lies in a proper connected $N_G(S)$-invariant subgroup D of G.

The constant $t(G)$ has been calculated, using a computer, by R. Lawther (private communication), in all cases except $G = E_8$: $t(G) = 12, 68, 124$ or 388, according as $G = G_2, F_4, E_6$ or E_7, respectively.

Theorem 5.10 covers all but finitely many simple subgroups S. If D is as in the conclusion, then of course $DN_G(S)$ lies in a maximal closed subgroup of G of positive dimension; such subgroups are determined by 5.6 if $p = 0$ or $p > 7$.

A number of consequences of 5.10, particularly for the finite groups G^F, are obtained in [LS4]. We mention just one:

Corollary 5.11 ([LS4]) *There is a constant c, such that if H is a maximal subgroup of the finite exceptional group G^F (F a Frobenius morphism), with $|H| > c$, then either*

(i) *H is the centralizer of a field, graph or graph-field automorphism of G^F, or*

(ii) *$H = \bar{X}^F$ for some maximal closed F-stable subgroup $\bar{X}$ of G of positive dimension.*

The subgroups under (i) are just subgroups of the same type as G^F, possibly twisted, over subfields of the defining finite field, and are determined up to conjugacy by [LS2, 5.1]; the subgroups under (ii) are given by 5.6, provided $p = 0$ or $p > 7$. As remarked after 5.6, work is currently under way on removing these characteristic restrictions.

References

[As] M. Aschbacher, "On the maximal subgroups of the finite classical groups", *Invent. Math.* **76** (1984), 469-514.

[ABS] H. Azad, M. Barry and G.M. Seitz, "On the structure of parabolic subgroups", *Comm. in Alg.* **18** (1990), 551-562.

[BT] A. Borel and J. Tits, "Éléments unipotents et sousgroupes paraboliques de groupes réductifs", *Invent. Math.* **12** (1971), 95-104.

[BS] A. Borel and J. de Siebenthal, "Les sous-groupes fermés de rang maximum des groupes de Lie clos", *Comment. Math. Helv.* **23** (1949), 200-221.

[Bor] A. Borovik, "The structure of finite subgroups of simple algebraic groups", *Algebra and Logic* **28** (1989), 249-279 (in Russian).

[Bou] N. Bourbaki, *Groupes et algèbres de Lie* (Chapters 4,5 and 6), Hermann, Paris, 1968.

[Ca1] R.W. Carter, "Conjugacy classes in the Weyl group", *Compositio Math.* **25** (1972), 1-59.

[Ca2] R.W. Carter, "Centralizers of semisimple elements in finite groups of Lie type", *Proc. London Math. Soc.* **37** (1978), 491-507.

[Ca3] R.W. Carter, *Finite groups of Lie type: conjugacy classes and characters*, Wiley Interscience, 1985.

[Ch] C. Chevalley, *Seminaire Chevalley, Vols. I,II: classifications des groupes de Lie algébriques*, Paris, 1956-8.

[CG] A.M. Cohen and R.L. Griess, "On finite simple subgroups of the complex Lie group of type E_8", *Proc. Symp. Pure Math.* **47** (1987), 367-405.

[CGL] A.M. Cohen, R.L. Griess and B. Lisser, "The group $L(2,61)$ embeds in the Lie group of type E_8", *Comm. in Alg.* **21** (1993), 1889-1907.

[CLSS] A.M. Cohen, M.W. Liebeck, J. Saxl and G.M. Seitz, "The local maximal subgroups of exceptional groups of Lie type, finite and algebraic", *Proc. London Math. Soc.* **64** (1992), 21-48.

[CW1] A.M. Cohen and D.B. Wales, "Finite subgroups of $G_2(\mathbb{C})$", *Comm. in Alg.* **11** (1983), 441-459.

[CW2] A.M. Cohen and D.B. Wales, "Finite subgroups of $E_6(\mathbb{C})$ and $F_4(\mathbb{C})$", *Proc. London Math. Soc.* **74** (1997), 105-150.

[Co1] B.N. Cooperstein, "Subgroups of exceptional groups of Lie type generated by long root elements, I. Odd characteristic", *J. Algebra* **70** (1981), 270-282.

[Co2] B.N. Cooperstein, "Subgroups of exceptional groups of Lie type generated by long root elements, II. Even characteristic", *J. Algebra* **70** (1981), 283-298.

[Co3] B.N. Cooperstein, "The geometry of root subgroups in exceptional groups II", *Geom. Dedicata* **15** (1983), 1-45.

[GL] D. Gorenstein and R. Lyons, "The local structure of finite groups of characteristic 2 type", *Mem. Amer. Math. Soc.* **42** (1983), No. 276, 1-731.

[GLMS] R.M. Guralnick, M.W. Liebeck, H.D. Macpherson and G.M. Seitz, "Modules for algebraic groups with finitely many orbits on subspaces", *J. Algebra* **196** (1997), 211-250.

[Ja] J.C. Jantzen, "Low dimensional representations of reductive groups are semisimple", in *Algebraic groups and Lie groups* (eds. G. Lehrer *et al.*), Austral. Math. Soc. Lecture Series **9** (1997), 255-266.

[Kac] V. Kac, "Some remarks on nilpotent orbits", *J. Algebra* **64** (1980), 190-213.

[Ka] W.M. Kantor, "Subgroups of classical groups generated by long root elements", *Trans. Amer. Math. Soc.* **248** (1979), 347-379.

[KR] P.B. Kleidman and A.J.E. Ryba, "Kostant's conjecture holds for E_7: $L_2(37) < E_7(\mathbb{C})$", *J. Algebra* **161** (1993), 535-540.

[LaSe] V. Landazuri and G.M. Seitz, "On the minimal degrees of projective representations of the finite Chevalley groups", *J. Algebra* **32** (1974), 418–443.

[LT] R. Lawther and D.M. Testerman, "A_1 subgroups of exceptional algebraic groups", *Trans. Amer. Math. Soc.*, to appear.

[LS1] M.W. Liebeck and G.M. Seitz, "Maximal subgroups of exceptional groups of Lie type, finite and algebraic", *Geom. Dedicata* **36** (1990), 353-387.

[LS2] M.W. Liebeck and G.M. Seitz, "Subgroups generated by root elements in groups of Lie type", *Annals of Math.* **139** (1994), 293-361.

[LS3] M.W. Liebeck and G.M. Seitz, "Reductive subgroups of exceptional algebraic groups", *Mem. Amer. Math. Soc.* **121** (1996), No. 580, 1-111

[LS4] M.W. Liebeck and G.M. Seitz, "On the subgroup structure of exceptional groups of Lie type", *Trans. Amer. Math. Soc.*, to appear

[LS5] M.W. Liebeck and G.M. Seitz, "On the subgroup structure of algebraic and finite classical groups", *Invent. Math.*, to appear.

[Lu] G. Lusztig, "On the finiteness of the number of unipotent classes", *Invent. Math.* **34** (1976), 201-213.

[RRS] R. Richardson, G. Röhrle and R. Steinberg, "Parabolic subgroups with abelian unipotent radical", *Invent. Math.* **110** (1992), 649-671.

[Rö] G. Röhrle, "On the structure of parabolic subgroups in algebraic groups", *J. Algebra* **157** (1993), 80-115.

[SaSe] J. Saxl and G.M. Seitz, "Subgroups of algebraic groups containing regular unipotent elements", *J. London Math. Soc.* **55** (1997), 370-386.

[Se1] G.M. Seitz, "The maximal subgroups of classical algebraic groups", *Mem. Amer. Math. Soc.* **67** (1987), No. 365, 1-286.

[Se2] G.M. Seitz, "Representations and maximal subgroups of finite groups of Lie type", *Geom. Dedicata* **25** (1988), 391-406.

[Se3] G.M. Seitz, "Maximal subgroups of exceptional algebraic groups", *Mem. Amer. Math. Soc.* **90** (1991), No. 441, 1-197.

[ST] G.M. Seitz and D.M. Testerman, "Extending morphisms from finite to algebraic groups", *J. Algebra* **131** (1990), 559-574.

[Ser] J-P. Serre, "Exemples de plongements des groupes $\mathrm{PSL}_2(F_p)$ dans des groupes de Lie simples", *Invent. Math.* **124** (1996), 525-562.

[Sl] P. Slodowy, "Two notes on a finiteness problem in the representation theory of finite groups", in *Algebraic groups and Lie groups* (eds. G. Lehrer *et al.*), Austral. Math. Soc. Lecture Series **9** (1997), 331-348.

[SS] T.A. Springer and R. Steinberg, "Conjugacy classes", in: *Seminar on algebraic groups and related topics* (eds. A. Borel *et al.*), Lecture Notes in Math. **131**, Springer, Berlin, 1970, pp. 168–266.

[St1] R. Steinberg, *Lectures on Chevalley groups*, Yale University Lecture Notes, 1968.

[St2] R. Steinberg, "Endomorphisms of linear algebraic groups", *Mem. Amer. Math. Soc.*, No. 80 (1968).

[Te] D.M. Testerman, "A_1-type overgroups of elements of order p in semisimple algebraic groups and the associated finite groups", *J. Algebra* **177** (1995), 34-76.

[We] A. Weil, "Remarks on the cohomology of groups", *Annals of Math.* **80** (1964), 149-157.

An Introduction to Intersection Cohomology

Jeremy Rickard

School of Mathematics, University of Bristol, University Walk, Bristol BS8 1TW, England

1 Introduction

Natural examples of singular varieties often arise in the study of algebraic groups. For example, if G is a connected reductive group with a Borel subgroup B, then B acts by left multiplication on the flag variety G/B with orbits BwB/B indexed by the elements $w \in W$ of the Weyl group. Although the variety BwB/B (called a **Bruhat cell**) is nonsingular, its closure in G/B (called a **Schubert cell**) is usually a singular variety.

Homology and cohomology have long been powerful tools for the study of complex algebraic varieties (and other topological spaces), and when ℓ-adic cohomology was introduced by Grothendieck to tackle the Weil conjectures, this provided a corresponding tool for the study of algebraic varieties over fields of prime characteristic.

However, ordinary (co)homology of manifolds and algebraic varieties works better when they are nonsingular. For example, many theorems and techniques such as Poincaré duality and Hodge theory do not work for singular varieties. In the early 1980's Goresky and MacPherson defined a new kind of homology, called **intersection homology**, which is identical to ordinary homology for nonsingular varieties, but is better for singular varieties since it *does* have desirable properties such as Poincaré duality. Since then this new tool, and developments of it such as ℓ-adic intersection cohomology, have been used to great effect in the study of algebraic groups, most notably in the work of Lusztig. In this article we aim to give a brief introduction to the theory and to describe very briefly one application: Kazhdan and Lusztig's interpretation of the coefficients of Kazhdan-Lusztig polynomials (see Donkin's lectures in this volume for a description of the importance of these in representation theory) as dimensions of intersection cohomology groups.

Our plan in this article is as follows. First we shall describe the original topological version of intersection homology, due to Goresky and MacPherson [3], after briefly reviewing ordinary homology theory. Many of the applications to algebraic groups require variations of the original theory (such as the ℓ-adic theory, if we wish to consider the characteristic $p > 0$ case), so we shall then go on to give a brief review of sheaf theory and describe the more powerful sheaf theoretic version of intersection cohomology, also due to Goresky and MacPherson [4]. We shall then describe a more algebraic interpretation, due to Deligne, which is vital when we then go on to sketch the definition of the ℓ-adic version of the theory. Finally we shall give a brief description of the interpretation of Kazhdan-Lusztig polynomials in terms of intersection cohomology.

A more detailed and extensive introduction to intersection homology, and a more comprehensive list of references up to the late 1980's can be found in Kirwan's book [6]. I would like to express my great debt to this book, which I found invaluable when preparing this article and the lectures on which it is based. Almost all the theorems on intersection homology described in this article were originally proved in the two fundamental papers [3, 4] of Goresky and MacPherson.

2 Simplicial homology

Let X be a topological space.

A **triangulation** of X is a homeomorphism

$$\mathcal{T} : |\Delta| \longrightarrow X,$$

where Δ is a simplicial complex and $|\Delta|$ is its geometric realization.

Let us fix, for the time being, the triangulation $\mathcal{T}$ of X; often we shall sloppily identify X with $|\Delta|$ via the homeomorphism $\mathcal{T}$.

Since the geometric realization of a single simplex is compact, X must be compact if Δ is finite, being a finite union of compact sets. We shall need to consider non-compact spaces X, so we shall not insist that $|\Delta|$ is finite, but only that it is **locally finite**, meaning that each vertex of Δ is contained in only finitely many simplices. The topology on $|\Delta|$ is induced from the usual topology on each simplex: i.e., a subset U of $|\Delta|$ is open if and only if its intersection $U \cap \sigma$ with each simplex σ is open in σ. If Δ is locally finite, then in fact it is easy to see that X is compact if and only if Δ is finite.

We shall often refer to simplices by listing their vertices. For example,

$$\sigma = \langle x_0, \dots, x_n \rangle$$

will denote the n-dimensional simplex (or 'n-simplex') with vertices $x_0, \dots, x_n$. An **orientation** of a simplex of dimension $n > 0$ is a choice of ordering of the vertices up to even permutations, so a simplex has two possible orientations. When it is relevant, we shall indicate which orientation we have chosen by the order in which we list the vertices.

Consider our space X with triangulation $\mathcal{T}$. We shall fix a choice of orientation for each simplex of Δ, and for each natural number i we shall let

$C_i^{\mathcal{T}}(X)$ be the complex vector space generated by elements $[\sigma] = [x_0, \ldots, x_i]$, one for each i-simplex $\sigma = \langle x_0, \ldots, x_i \rangle$ of Δ. If $\tau = \langle y_0, \ldots, y_i \rangle$ is the i-simplex σ of Δ with the *opposite* orientation, we define $[\tau] = [y_0, \ldots, y_i]$ to be the element $-[\sigma]$ of $C_i^{\mathcal{T}}(X)$. We call $C_i^{\mathcal{T}}(X)$ the space of **simplicial i-chains** of X.

For any simplicial i-chain $\xi = \sum_\sigma \lambda_\sigma [\sigma]$, we can define the **support** of ξ to be the union

$$|\xi| = \bigcup_{\lambda_\sigma \neq 0} \sigma$$

of the i-simplices that occur in ξ with non-zero coefficient. This is a compact subset of X.

There is a linear map

$$\partial : C_i^{\mathcal{T}}(X) \longrightarrow C_{i-1}^{\mathcal{T}}(X)$$

for each $i > 0$ defined by

$$\partial([x_0, \ldots, x_i]) = [x_1, x_2, \ldots, x_i] - [x_0, x_2, \ldots, x_i] + \ldots + (-1)^i [x_0, \ldots, x_{i-2}, x_{i-1}].$$

It is easy to check that $\partial^2 = 0$, so that $(C_*^{\mathcal{T}}, \partial)$ is a chain complex of vector spaces whose homology is called the **simplicial homology** of X, with coefficients in $\mathbb{C}$, with respect to the triangulation $\mathcal{T}$. Up to natural isomorphism, the simplicial homology is well-known to be independent of the triangulation $\mathcal{T}$ and the choice of orientation of the simplices, so we shall denote it by $H_*(X)$, suppressing the triangulation, the orientations and the coefficients $\mathbb{C}$ from the notation for simplicity.

As a variation on this definition, we can define $C_i^{BM,\mathcal{T}}(X)$, the space of simplicial **Borel-Moore i-chains** of X, to be the complex vector space of all (possibly infinite) formal linear combinations

$$\sum_\sigma \lambda_\sigma [\sigma]$$

of the elements $[\sigma]$, one for each i-simplex of Δ. We can define the support $|\xi|$ of a Borel-Moore i-chain as before; in general it will be a closed, but not necessarily compact, subset of X. The formula for the differential ∂ still makes sense, since the triangulation is locally finite, and so for each $(i-1)$-simplex σ', $[\sigma']$ is only involved in $\partial([\sigma])$ for finitely many i-simplices σ. Hence we get a chain complex $(C_*^{BM,\mathcal{T}}(X), \partial)$, whose homology $H_*^{BM}(X)$ (again independent of the triangulation) is called the simplicial **Borel-Moore homology** (or homology with closed support) of X.

If X is compact, then Δ has only finitely many simplices, and so the Borel-Moore homology is precisely the same as the ordinary homology.

Since we have been dealing, for simplicity, with homology with coefficients in the *field* $\mathbb{C}$, it is now very easy to define the **cohomology** of X as the vector space dual of its homology. More precisely, the ordinary cohomology $H^*(X)$ is the dual of $H_*(X)$ and the **cohomology with compact support** $H_c^*(X)$ is the dual of $H_*^{BM}(X)$.

It is well-known that ordinary homology is functorial: a continuous map $f : X \longrightarrow Y$ induces a linear map $f_i : H_i(X) \longrightarrow H_i(Y)$ for each i. For the simplicial version of homology that we have described, this is only transparent if the map f is simplicial; for general continuous maps it is clearer if we use some other version of homology, such as 'singular homology'. Since cohomology is dual to homology, this is also functorial, but this time it is a contravariant functor. Also, ordinary homology (and cohomology) are **homotopy invariants**: if X and Y are homotopy equivalent, then $H_*(X)$ is isomorphic to $H_*(Y)$.

These facts are not true in general for Borel-Moore homology. For example, if X is the real line, then we can triangulate it in an obvious way, and it is easy to calculate that $H_i^{BM}(X) \cong 0$ for $i \neq 1$ and $H_1^{BM}(X) \cong \mathbb{C}$. However, if Y is a single point, then $H_i^{BM}(Y) \cong 0$ for $i \neq 0$ and $H_0^{BM}(Y) \cong \mathbb{C}$. Hence X and Y have different Borel-Moore homology, even though they are homotopy equivalent. This example also shows that Borel-Moore homology can't be functorial, since the identity map of Y factors through X, whereas the identity map of $H_0^{BM}(Y) \cong \mathbb{C}$ certainly can't factor through $H_0^{BM}(X) \cong 0$.

Although we shall not investigate in detail the fact that simplicial homology is independent of the choice of triangulation, it will be convenient to have a description that at least makes it clear that replacing one triangulation with a certain kind of finer triangulation will not make any difference.

If $\mathcal{T}' : |\Delta'| \longrightarrow X$ is another triangulation of X, we shall say that $\mathcal{T}'$ is a **linear refinement** of $\mathcal{T}$ if the resulting homeomorphism from $|\Delta'|$ to $|\Delta|$ maps each simplex of Δ' to a subset of some simplex of Δ via a *linear* map. Identifying the three spaces X, $|\Delta|$ and $|\Delta'|$ via the homeomorphisms, we shall often sloppily say that a simplex of Δ' *is* a subset of a simplex of Δ.

More generally, if $U \subseteq X$ then we can similarly define a triangulation

$$\mathcal{S} : |\Sigma| \longrightarrow U$$

to be a **linear refinement** of the triangulation $\mathcal{T}$ of X if it induces linear maps from the simplices of Σ to subsets of the simplices of Δ. As above, we shall often identify the simplices of Σ with the corresponding subsets of the simplices of Δ. If U is an open subset of X then it always possible to triangulate U by a linear refinement of $\mathcal{T}$, although of course this will not usually be a finite triangulation, even if $\mathcal{T}$ is finite.

We shall formally define a **piecewise linear structure** on a space X to be an equivalence class of triangulations of X, where two triangulations are equivalent if they have a common linear refinement (this is an equivalence relation because any two linear refinements of a given triangulation do have a common linear refinement). Thus any triangulation of X defines a piecewise linear structure on X, and any linear refinement defines the same piecewise-linear structure.

Also, a piecewise linear structure on X (represented by a triangulation $\mathcal{T}$, say) induces a piecewise linear structure on any open subset U of X, represented by any triangulation of U that is a linear refinement of $\mathcal{T}$.

Now if $\mathcal{T}'$ is a linear refinement of $\mathcal{T}$ and σ' is an i-simplex of Δ' that is contained in an i-simplex of Δ, then an orientation of σ determines an orientation of σ'. [To see this, pick an interior point of σ' and consider the projection of the boundary of σ' to the boundary of σ from this point. Given orderings of the vertices of the two simplices this determines a homeomorphism from the boundary B of a standard i-simplex to itself. This will induce the identity map from the degree $i-1$ homology $H_{i-1}(B)$ to itself if and only if the chosen orientation of σ' is the one determined by the chosen orientation of σ.] If we have fixed an orientation of each simplex of Δ, we shall call a choice of orientations for the simplices of Δ' **compatible** if the orientations are determined by those of Δ whenever possible. Note that there is not a unique compatible choice of orientations: some i-simplices of Δ' are only contained in simplices of Δ of larger dimension, and for these we are free to choose the orientation however we wish.

Given a compatible choice of orientations for the simplices of Δ and Δ', there is a linear map from $C_i^{\mathcal{T}}(X)$ to $C_i^{\mathcal{T}'}(X)$ mapping

$$[\sigma] \mapsto \sum_{\sigma' \in \Delta',\, \sigma' \subseteq \sigma} [\sigma'],$$

the sum being over all i-simplices σ' of Δ' contained in an i-simplex σ of Δ. It is easy to check that this commutes with the differential, and so it induces a chain map

$$(C_*^{\mathcal{T}}, \partial) \longrightarrow (C_*^{\mathcal{T}'}, \partial).$$

This chain map induces an isomorphism on homology.

We can now define a notion of i-chain that depends only on the choice of piecewise linear structure on X, not on the particular triangulation. To do this we just define $C_i(X)$, the space of **piecewise linear i-chains** of X, to be

$$C_i(X) = \lim_{\longrightarrow} C_i^{\mathcal{T}'}(X),$$

the direct limit being taken over *all* linear refinements $\mathcal{T}'$ of $\mathcal{T}$. Thus an i-simplex τ of any linear refinement of $\mathcal{T}$ determines an element $[\tau]$ of $C_i(X)$. By the discussion above, we get a chain complex $(C_*(X), \partial)$ whose homology is naturally isomorphic to $H_*^{\mathcal{T}}(X)$, since homology commutes with taking direct limits for complexes of abelian groups.

Similarly we can define the piecewise linear Borel-Moore i-chains of X, obtaining a complex $(C_*^{BM}(X), \partial)$ that depends only on the choice of a piecewise linear structure on X.

Now let U be an open subset of X with a triangulation $\mathcal{S} : |\Sigma| \longrightarrow U$ that is a linear refinement of $\mathcal{T}$. There is no obvious way to define a map from $C_i^{\mathcal{T}}(X)$ to $C_i^{\mathcal{S}}(U)$, since an i-simplex of Δ can contain an infinite number of i-simplices of Σ. However, if we work with Borel-Moore i-chains this is no longer a problem, and we do get a chain map

$$(C_*^{BM,\mathcal{T}}(X), \partial) \longrightarrow (C_*^{BM,\mathcal{S}}(U), \partial)$$

by mapping

$$[\sigma] \mapsto \sum_{\sigma' \in \Sigma,\ \sigma' \subseteq \sigma} [\sigma'],$$

the (formal) sum being over all i-simplices σ' of Σ contained in an i-simplex σ of Δ. If we now take the direct limit over all linear refinements of $\mathcal{T}$ and $\mathcal{S}$, we get a chain map

$$(C_*^{BM}(X), \partial) \longrightarrow (C_*^{BM}(U), \partial).$$

Although we have omitted it from the notation for the sake of tidyness, we should emphasize that there is a choice of piecewise linear structure underlying this.

Let us finish this section by briefly recalling the statement of Poincaré duality. Let X be a compact n-dimensional manifold (for example, if $n = 2d$ is even, X could be a non-singular d-dimensional projective variety over $\mathbb{C}$), with a triangulation $\mathcal{T} : |\Delta| \longrightarrow X$. Since X is compact, Δ has only finitely many simplices, so the sum

$$\overline{X} = \sum_{\sigma \in \Delta,\ \dim(\sigma)=n} [\sigma]$$

is an element of $C_n^{\mathcal{T}}(X)$. If there is a choice of orientations of the simplices of Δ for which $\overline{X} \in \ker(\partial)$, then X is said to be **orientable** (for example, this is the case for a sphere, but not for a Klein bottle): in this case $\overline{X}$ determines an element of $H_n(X)$, which we shall also call $\overline{X}$, which, if X is connected, is independent (up to a sign) of the choice of triangulation and orientation.

Poincaré duality states that for each i there is a natural non-degenerate pairing

$$H_i(X) \times H_{n-i}(X) \longrightarrow \mathbb{C},$$

or equivalently an isomorphism between $H_i(X)$ and the dual $H^{n-i}(X)$ of $H_{n-i}(X)$. In particular this implies that $H_i(X)$ and $H_{n-i}(X)$ have the same dimension for each i. This pairing has a geometric interpretation in terms of intersections of submanifolds that we shall now sketch in the case that X is connected.

Let M and N be closed orientable submanifolds of X, with dimensions i and $n - i$ respectively. If M and N are 'in general position' then their intersection $L = M \cap N$ is just a finite subset of X. The inclusion maps from M, N and L to X take the elements $\overline{M} \in H_i(M)$, $\overline{N} \in H_{n-i}(N)$ and $\overline{L} \in H_0(L)$ to elements $m \in H_i(X)$, $n \in H_{n-i}(X)$ and $l \in H_0(X)$. With a suitable choice of orientations (this is a little subtle for the zero-dimensional manifold L), the Poincaré duality pairing takes (m, n) to $l \in H_0(X) \cong \mathbb{C}$.

This requires X to be a genuine manifold (i.e., not a 'manifold with singularities'). This can be seen from a simple example. Let X be the union of two n-dimensional spheres intersecting in a single point (where $n > 0$); if we remove this one singular point, then X becomes an n-dimensional manifold. However, the homology of X is

$$H_i(X) \cong \begin{cases} \mathbb{C} & \text{if } i = 0 \\ \mathbb{C}^2 & \text{if } i = n \\ 0 & \text{if } i \neq 0, n, \end{cases}$$

and so Poincaré duality can't hold, since $H_0(X)$ and $H_n(X)$ have different dimensions, and so can't possibly be dual vector spaces.

3 Simplicial intersection homology

To define intersection homology, we shall need to restrict the kind of topological space that we consider.

Let us define an n-dimensional triangulated **pseudomanifold** to be a space X with a triangulation $\mathcal{T} : |\Delta| \longrightarrow X$ and a closed subspace $\Sigma \subseteq X$ that is a union of simplices, such that $X - \Sigma$ is an n-dimensional manifold that is dense in X and such that

$$\dim(\Sigma) \leq \dim(X) - 2.$$

For example, if X is a triangulated manifold then we can make X into a pseudomanifold by taking $\Sigma = \emptyset$.

Another important class of examples is given by varieties over the complex numbers. If X is any complex quasi-projective variety of pure dimension d (i.e., every irreducible component of X is d-dimensional), then X can be made into a $2d$-dimensional triangulated pseudomanifold, taking Σ to be the singular locus of X. See Sections 3.2 and 3.3 of [6] for references to the classical theorems on stratifications and triangulations that imply this.

Given an n-dimensional triangulated pseudomanifold X as above, we can define a stratification

$$X_0 \subseteq X_1 \subseteq \ldots \subseteq X_{n-2} \subseteq X_n = X$$

of X by taking $X_{n-2} = \Sigma$ and taking X_i to be the union of all the i-dimensional simplices of Σ for $i < n - 2$.

Various other stratifications of X will do, but we shall not go into the technical details of the topological conditions they have to satisfy. One important fact is that if X is a complex variety, it is possible to choose the stratification so that $X_{2j+1} = X_{2j}$ is a closed j-dimensional subvariety of X for each j. Also, when we come to take a linear refinement of $\mathcal{T}$, the original stratification will still do: we won't need to add all of the extra i-simplices to each stratum X_i. Whatever stratification we use, one of the conditions needed is that $X_k - X_{k-1}$ should be a k-dimensional manifold for each k.

The definition of intersection homology depends not just on X (and its triangulation and stratification) but also on the choice of a finite sequence of integers. By definition, a **perversity** is a finite sequence $\overline{p} = (p_2, \ldots, p_n)$ of integers such that $p_2 = 0$ and for each j, either $p_{j+1} = p_j$ or $p_{j+1} = p_j + 1$.

The extreme cases are thus the **top** perversity $\overline{t} = (0, 1, \ldots, n-2)$ and the **bottom** perversity $\overline{0} = (0, 0, \ldots, 0)$. If $\overline{p}$ is any perversity, then so is $\overline{t} - \overline{p} = (0, 1-p_3, 2-p_4, \ldots, n-2-p_n)$; this is called the **complementary** perversity of $\overline{p}$. For example, the top and bottom perversities are complementary to one another.

Given X, triangulated and stratified as above, and a choice of perversity $\overline{p}$, we define a subspace $IC_i^{\overline{p},\mathcal{T}}(X)$ of the space $C_i^{\mathcal{T}}(X)$ of simplicial i-chains by taking the set of all $\xi \in C_i^{\mathcal{T}}(X)$ such that:

- $\dim(|\xi| \cap X_{n-k}) \leq i - k + p_k$ for all k.
- $\dim(|\partial\xi| \cap X_{n-k}) \leq i - k + p_k - 1$ for all k.

We shall call $IC_i^{\overline{p},\mathcal{T}}(X)$ the space of simplicial **intersection** i**-chains** of X with respect to the chosen stratification and perversity.

Let us make some elementary remarks on the definition.

- For any $\overline{p}$, the first inequality (for $k = 2$) says
  $$\dim(|\xi| \cap \Sigma) \leq i - 2 < i = \dim(|\xi|),$$
  so ξ can't involve any of the i-simplices of Σ.
- If we use a stratification for which $X_{n-k} = X_{n-k-1}$, then the value of p_k is irrelevant.
- The second inequality ensures that
  $$\partial(IC_i^{\overline{p},\mathcal{T}}(X)) \subseteq IC_{i-1}^{\overline{p},\mathcal{T}}(X),$$
  and so $(IC_*^{\overline{p},\mathcal{T}}(X), \partial)$ is a subcomplex of $(C_*^{\mathcal{T}}(X), \partial)$.
- If $\Sigma = \emptyset$, then both inequalities are vacuous, so
  $$(IC_*^{\overline{p},\mathcal{T}}(X), \partial) = (C_*^{\mathcal{T}}(X), \partial).$$

- There is no problem in defining the dimension of $|\xi| \cap X_{n-k}$, since it is a union of simplices. Those who are familiar with the definition of singular homology may have wondered why we did not use this (more elegant) definition in the previous section. The reason is that for a *singular* i-chain ξ there is no reasonable definition of this dimension.

As a consequence of the third remark, we can define the simplicial **intersection homology** $IH_*^{\overline{p},\mathcal{T}}(X)$ to be the homology of $IC_*^{\overline{p},\mathcal{T}}(X)$. By the fourth remark this will agree with ordinary simplicial homology if $\Sigma = \emptyset$. It also has the following important properties, which we shall not prove here.

- It is independent of the triangulation.
- It is independent of the stratification (subject to the various topological restrictions on the stratification that that we have not made explicit).
- It is even independent of the choice of subspace Σ.

If X is a d-dimensional complex variety (so $n = 2d$) and we use a stratification by closed subvarieties, so that $X_{2j+1} = X_{2j}$, then (as we remarked above) the value of p_k is irrelevant for k odd. If we choose the perversity $\overline{m} = (0, 0, 1, 1, \ldots, d-2, d-2, d-1)$, called the **middle perversity**, then the complementary perversity $\overline{t} - \overline{m} = (0, 1, 1, 2, \ldots, d-2, d-1, d-1)$ agrees with $\overline{m}$ in all even places. In other words, the middle perversity is self-complementary for all practical purposes if we are dealing with a complex variety.

As a very simple example, let us calculate the degree zero intersection homology of a general pseudomanifold X.

Let $|\xi| \in C_i^{\mathcal{T}}(X)$. The condition for $|\xi|$ to be an intersection i-chain is just that

$$\dim(|\xi| \cap X_{n-k}) \leq i - k + p_k$$

for each k. But since $p_k - k \leq -2$, whatever the perversity, the condition is equivalent, for $i = 0$ or $i = 1$, to

$$|\xi| \cap \Sigma = \emptyset.$$

It follows that $IH_0^{\overline{p},\mathcal{T}}(X)$ has a natural basis indexed by the path components of $X - \Sigma$.

As with ordinary simplicial homology, there are several variations of the definition. We shall not give the definitions, as they are completely straightforward generalizations of what has gone before, but let us run through some of the possibilities.

- We can define the **Borel-Moore intersection homology** groups $IH_*^{BM,\overline{p},\mathcal{T}}(X)$ of X. In the literature this is sometimes referred to simply as 'intersection homology': for X compact, of course, it is the same as what we have called intersection homology. An easy calculation similar to the one above shows that $IH_0^{BM,\overline{p},\mathcal{T}}(X)$ has a basis given by those path components of $X - \Sigma$ that have compact closure in X.

- We can define the **intersection cohomology** $IH^*_{\overline{p},\mathcal{T}}(X)$ as the vector space dual of the intersection homology. Similarly the **intersection cohomology with compact support** $IH^*_{\overline{p},\mathcal{T},c}$ is dual to Borel-Moore intersection homology.

- Given a piecewise linear structure on X that is compatible with the stratification (e.g., the piecewise linear structure defined by the original triangulation), we can define 'piecewise linear intersection i-chains' and 'piecewise linear Borel-Moore intersection i-chains' in the same way as we did for ordinary homology, thus obtaining chain complexes that depend only on the choice of piecewise linear structure, and not on the particular triangulation.

- Given an open subset $U \subseteq X$, U inherits a piecewise linear structure and a stratification from those of X (the stratification is given by taking $U_k = U \cap X_k$). As in the case of ordinary homology, there is a natural chain map
$$(IC^{\overline{p}}_*(X), \partial) \longrightarrow (IC^{\overline{p}}_*(U), \partial)$$
between the piecewise linear intersection chain complexes of X and U.

Next we shall give a non-trivial fact that allows us to calculate intersection homology quite easily in certain cases, and will allow us to deduce that some properties of ordinary homology do *not* generalize.

We shall define X to be **topologically normal** if every point $x \in X$ has arbitrarily small neighbourhoods U such that $U - \Sigma$ is connected. For example, any (non-singular) manifold is normal, since we can take $\Sigma = \emptyset$.

The reason for the terminology is that if X is a complex variety that is normal in the sense of algebraic geometry, then it follows from Zariski's Main Theorem [5, Theorem V.5.2] that it is topologically normal.

A **normalization** of an n-dimensional pseudomanifold X is a surjective continuous map
$$\pi : \tilde{X} \longrightarrow X$$
from a topologically normal n-dimensional pseudomanifold $\tilde{X}$ such that π restricts to a homeomorphism
$$\pi^{-1}(X - \Sigma) \cong X - \Sigma$$
and such that $\pi^{-1}(x)$ is finite for each point $x \in X$.

A useful theorem of Goresky and MacPherson then states that the intersection homology of $\tilde{X}$ is naturally isomorphic to that of X.

For example, let X be the union of two n-dimensional spheres intersecting in a single point (we used this space earlier as an example to show that Poincaré duality does not hold in general for manifolds with singularities). The disjoint union $\tilde{X}$ of two n-dimensional spheres is a normalization of X, and in this case $\tilde{X}$ is a non-singular manifold, so its intersection homology coincides with its ordinary homology. We therefore deduce that

$$IH_i^{\overline{p}}(X) \cong H_i(\tilde{X}) \cong \begin{cases} \mathbb{C}^2 & \text{if } i = 0, n \\ 0 & \text{otherwise,} \end{cases}$$

independently of the perversity.

As the previous example illustrates, if X has a non-singular normalization, then the intersection homology of X will not depend on $\overline{p}$; for a general X, however, the choice of perversity is more important, as is demonstrated by the following results of Goresky and MacPherson calculating the intersection homology of a topologically normal X with respect to the top and bottom perversities.

- If X is a topologically normal n-dimensional pseudomanifold, then

$$IH_i^{\overline{0}}(X) \cong H^{n-i}(X).$$

- If X is a topologically normal n-dimensional pseudomanifold, then

$$IH_i^{\overline{t}}(X) \cong H_i(X).$$

Since $H^{n-i}(X)$ is a contravariant functor of X but $H_i(X)$ is a covariant functor, these results strongly suggest that intersection homology will not be functorial in general. This can be verified by the following simple example.

Let X be the union of two planes intersecting in two points. Then X is a 2-dimensional pseudomanifold with a non-singular normalization $\tilde{X}$ that is the disjoint union of two planes. Hence, for any perversity $\overline{p}$,

$$IH_i^{\overline{p}}(X) \cong H_i(\tilde{X}) \cong \begin{cases} \mathbb{C}^2 & \text{if } i = 0 \\ 0 & \text{otherwise.} \end{cases}$$

However, it is easy to see that X has a circle Y as a retract, and

$$IH_i^{\overline{p}}(Y) \cong H_i(Y) \cong \begin{cases} \mathbb{C} & \text{if } i = 0, 1 \\ 0 & \text{otherwise,} \end{cases}$$

since Y is a (non-singular) 1-dimensional manifold. If $IH_1^{\overline{p}}(X)$ were a functor in X (either contravariant or covariant), then $IH_1^{\overline{p}}(Y)$ would have to be a retract of $IH_1^{\overline{p}}(X)$, which is not the case.

In the last example, X and Y are homotopy equivalent, so the same example demonstrates that intersection homology is not a homotopy invariant.

We hinted earlier that intersection homology would allow us to recover Poincaré duality for singular varieties. We shall end this section with the precise statement, again due to Goresky and MacPherson.

Let X be a compact n-dimensional pseudomanifold. Let $\overline{p}$ and $\overline{q}$ be a complementary pair of perversities. Then there is a nondegenerate pairing

$$IH_i^{\overline{p}}(X) \times IH_{n-i}^{\overline{q}}(X) \longrightarrow \mathbb{C}$$

for each i with $0 \leq i \leq n$.

This is slightly unsatisfactory, because it involves two different perversities. However, if we take $\overline{p}$ to be the middle perversity $\overline{m}$, then we have seen that the intersection homology of a complex variety with respect to $\overline{m}$ is the same as that with respect to the complementary perversity $\overline{t} - \overline{m}$, so in this case the statement becomes more satisfactory: if X is a projective complex variety of pure (complex) dimension d then there is a nondegenerate pairing

$$IH_i^{\overline{m}}(X) \times IH_{2d-i}^{\overline{m}}(X) \longrightarrow \mathbb{C},$$

and so $IH_i^{\overline{m}}(X)$ is naturally dual to $IH_{2d-i}^{\overline{m}}(X)$. The condition that X should be a projective variety is to ensure that it is compact as a pseudomanifold

For this and other reasons, the middle perversity is the natural one to use for complex varieties, and the notation is often simplified to remove the reference to $\overline{m}$, defining

$$IH_i(X) = IH_i^{\overline{m}}(X).$$

4 Sheaf theory

In this section we shall give a brief review of the basics of the theory of sheaves and their cohomology. This theory gives a more algebraic way of defining the cohomology of a topological space than the simplicial theory we previously described, and allows the machinery of derived functors and derived categories (see Keller's lectures in this volume) to be brought to bear. It also allows generalizations of the topological theory, such as étale and ℓ-adic cohomology, which we shall very briefly mention in the second half of this section.

4.1 Topological sheaf theory

Before defining sheaves, we shall start with the more elementary notion of presheaves. A **presheaf** F of abelian groups on a topological space X consists of the following data:

- An abelian group $F(U)$ for each open subset U of X.

- A homomorphism $\rho^U_V : F(U) \longrightarrow F(V)$, called a **restriction map**, for each pair of open subsets $V \subseteq U$ of X, such that:

  1. ρ^U_U is the identity map on $F(U)$ for each U.
  2. $\rho^V_W \rho^U_V = \rho^U_W$ whenever $W \subseteq V \subseteq U$.

Similarly we can define a presheaf of vector spaces, or indeed a presheaf of objects from any category. The elements of $F(U)$ are called **sections** of F over U, and elements of $F(X)$ are called **global sections**.

We can regard the set of open sets of a space X as the objects of a category, where the morphisms are just the inclusion maps (so there is at most one morphism between any two objects). If we focus on this category, the definition of a presheaf can be rephrased in a simple category theoretic form: a presheaf of abelian groups on X is just a contravariant functor F from the category of open sets of X to the category of abelian groups.

Given an abelian group A, one very simple example of a presheaf on X is the **constant presheaf** A. Here $A(U) = A$ for every U, and ρ^U_V is the identity map for all U and $V \subseteq U$.

A **homomorphism** $\alpha : F \longrightarrow G$ of presheaves is just a natural transformation of functors if we interpret presheaves as functors. More explicitly, it consists of a family of homomorphisms $\alpha_U : F(U) \longrightarrow G(U)$, one for each open subset U of X, that commute with the restriction maps.

Let F be a presheaf on X and let $x \in X$. We can define an abelian group F_x, called the **stalk** of F at x: this is just

$$F_x = \varinjlim F(U),$$

the direct limit taken over all open subsets $U \subseteq X$ with $x \in U$. More concretely, every element of the stalk is represented by a section of F over some open neighbourhood of x, with two such sections representing the same element of F_x if they agree on restriction to some smaller neighbourhood of x.

A sheaf is a special kind of presheaf: the general idea is that a sheaf is a presheaf where the sections are 'determined locally'. To be precise, a **sheaf** F of abelian groups on X is a presheaf satisfying the following conditions:

- If $\{U_i : i \in I\}$ is an open cover of U and $s \in F(U)$, then $s = 0$ if and only if $\rho^U_{U_i}(s) = 0$ for every $i \in I$.

- If $\{U_i : i \in I\}$ is an open cover of U and for each $i \in I$ we choose $s_i \in F(U_i)$ so that for each pair (i, j),

$$\rho^{U_i}_{(U_i \cap U_j)}(s_i) = \rho^{U_j}_{(U_i \cap U_j)}(s_j),$$

then there is a section $s \in F(U)$ such that

$$\rho^U_{U_i}(s) = s_i$$

for every $i \in I$.

Note that the first condition implies that the section s of the second condition is *uniquely* determined by the s_i. Note also that if $U = \emptyset$ is the empty set, then we can choose an empty open cover of U, so the first condition implies that $F(\emptyset) = 0$ for a sheaf F.

One basic example of a sheaf is given by continuous functions: if we take $F(U)$ to be the set of all continuous functions from U to $\mathbb{R}$, with the usual notion of restriction of functions, then F is a sheaf of real vector spaces on X.

Notice that a constant presheaf A is *not* usually a sheaf: let U and V be disjoint open sets, so $\{U, V\}$ is an open cover of $U \cup V$. Then if we choose $s \in F(U) = A$ and $t \in F(V) = A$ to be *different* elements of A, then there is no element of $F(U \cup V)$ that restricts to both s and t, contradicting the second property of a sheaf. However, there is a related sheaf, called the **constant sheaf** A: we let $A(U)$ be the set of all locally constant functions from U to A. The sections of the constant sheaf and the constant presheaf only agree over *connected* open sets.

Going from the constant presheaf to the constant sheaf is an example of a general procedure, with the ugly name of **sheafification**, that turns presheaves into sheaves. Given any presheaf F, there is a sheaf F^+ and a presheaf homomorphism $F \longrightarrow F^+$ with the universal property that *any* map from F to a sheaf factors uniquely through $F \longrightarrow F^+$. Clearly then, if F is already a sheaf then the sheafification F^+ is just F itself. In terms of adjoint functors, sheafification is left adjoint to the inclusion of the category of sheaves into the category of presheaves. This procedure preserves all stalks: the natural map $F_x \longrightarrow F_x^+$ is an isomorphism for every point $x \in X$.

Often we want to emphasize the dependence of the group $F(U)$ of sections on the sheaf F rather than on the open set U. To do this we introduce an alternative notation: $\Gamma(U, F)$ is by definition just $F(U)$. So $\Gamma(U, ?)$ is a functor from the category of sheaves to the category of abelian groups.

A variation on $\Gamma(U, F)$ is $\Gamma_c(U, F)$, the group of **sections with compact support** of F over U. This is just the set of $s \in \Gamma(U, F)$ for which there is a compact subset $C \subseteq U$ so that $\rho^U_V(s) = 0$ whenever $V \subseteq U$ is an open subset with $V \cap C = \emptyset$. Of course, if $U = X$ and X is compact, then this condition is vacuous and so $\Gamma(X, ?) = \Gamma_c(X, ?)$.

It is clear that the category of presheaves of abelian groups on X forms an abelian category, since it is just a category of functors taking values in an abelian category. A more subtle fact is that the category of sheaves is also abelian. The reason that this is not trivial is that if $\alpha : F \longrightarrow G$ is a homomorphism of presheaves, then the cokernel of α is not necessarily a sheaf, even if both F and G are. However, the sheafification functor allows us to put this right: the cokernel, *in the category of sheaves*, of a map $\alpha : F \longrightarrow G$ of sheaves is just the sheafification of the cokernel, *in the category of presheaves*, of α.

There is no corresponding problem for kernels of maps: the kernel of a map $\alpha : F \longrightarrow G$ of presheaves is already a sheaf if both F and G are.

The subtlety surrounding cokernels has the following important consequence. The functor $\Gamma(U, ?)$, regarded as a functor from presheaves to abelian groups, is clearly exact. However, considered as a functor from *sheaves* to abelian groups it is not necessarily exact, since if

$$0 \longrightarrow F_1 \longrightarrow F_2 \longrightarrow F_3 \longrightarrow 0$$

is a short exact sequence of sheaves, F_3 is not necessarily isomorphic to the presheaf quotient F_2/F_1, and so

$$0 \longrightarrow F_1(U) \longrightarrow F_2(U) \longrightarrow F_3(U) \longrightarrow 0$$

is not necessarily exact at $F_3(U)$. Since the problem only occurs at $F_3(U)$, $\Gamma(U, ?)$ is a left exact functor: the sequence

$$0 \longrightarrow F_1(U) \longrightarrow F_2(U) \longrightarrow F_3(U)$$

is exact. This failure of exactness is the key to the importance of sheaf cohomology.

Let $\mathrm{Sh}(X)$ be the category of sheaves of abelian groups on X. As we have already said, $\mathrm{Sh}(X)$ is an abelian category, albeit in a slightly subtle way. It also has suitable properties for the application of homological algebra, in that it has enough injectives: every sheaf can be embedded in an injective object of $\mathrm{Sh}(X)$. This allows us to define the derived category of the category of sheaves and to define the right derived functors of any left exact functor from $\mathrm{Sh}(X)$ to another abelian category: see Keller's lectures for the generalities of these constructions.

Except in the most trivial cases, $\mathrm{Sh}(X)$ does *not* have enough projectives, so the construction of left derived functors is not so straightforward.

The most important example of a left exact functor from $\mathrm{Sh}(X)$ to abelian groups is the global section functor $\Gamma(X, ?)$. The right derived functors of this are called the **sheaf cohomology** of X: $H^i(X, F)$ is the i-th derived functor of $\Gamma(X, ?)$ applied to F. This gives a good algebraic alternative to simplicial or singular cohomology, since if we take F to be the constant sheaf $\mathbb{C}$, then for all reasonably nice spaces X the sheaf cohomology $H^*(X, \mathbb{C})$ is naturally isomorphic to the ordinary cohomology of X with coefficients in $\mathbb{C}$.

Similarly, if we define $H^i_c(X, ?)$ to be the i-th right derived functor of $\Gamma_c(X, ?)$ (global sections with compact support), then $H^*_c(X, \mathbb{C})$ agrees with ordinary cohomology with compact support, assuming X is reasonably nice.

We shall end this survey of sheaf theory by describing some examples of functors between categories of sheaves.

First, let $\beta : X \longrightarrow Y$ be a continuous map between topological spaces. If F is a sheaf on X, then we can define a sheaf $\beta_* F$ on Y by

$$\beta_* F(U) = F(\beta^{-1}(U))$$

for each open subset U of Y. This defines a functor β_*, called the **direct image** functor, from $\mathrm{Sh}(X)$ to $\mathrm{Sh}(Y)$. In general β_* is left exact but not exact, as is illustrated by the case where $Y = \{y\}$ has a single point. Then $\mathrm{Sh}(Y)$ is isomorphic to the category of abelian groups, and β_* is essentially just the global section functor $\Gamma(X, ?)$. We can, of course, form the right derived functors of β_* for any continuous map β.

Secondly, let $W \subseteq X$ be a subspace of X. If W is open in X, there is an obvious way of producing a sheaf $F|_W$ on W from a sheaf F on X, just by setting

$$F|_W(U) = F(U)$$

for an open subset U of W. Even if W is not open, we can define a presheaf F' on W by

$$F'(U) = \varinjlim F(V),$$

the direct limit being taken over all open subsets $V \subseteq X$ with $U \subseteq V$. This presheaf F' is not generally a sheaf, but we can sheafify it to produce a sheaf $F|_W$ on W, called the **restriction** of F to W. The functor $F \mapsto F|_W$ from $\mathrm{Sh}(X)$ to $\mathrm{Sh}(W)$ is always exact. Restriction generalizes the idea of stalks, since if $W = \{w\}$ has a single point, then it is immediate from the definitions that $\Gamma(W, F|_W)$ is just the stalk F_w of F at w.

4.2 Étale and ℓ-adic sheaves

As we tried to emphasize, the notion of a presheaf on a space X uses only the category of open sets of X, not the collection of points of X. Even when we go on to define sheaves, the only extra ingredients needed are the notions of open covers and of intersections. Of these, only the notion of an open cover is really an extra structure on the category of open sets of X, since the intersection of open sets U and V can be characterized by a universal property as the categorical product (in the category of open sets of X) of U and V. Grothendieck and his school of algebraic geometers realized that the whole of sheaf theory, including sheaf cohomology, could be carried out in the much more general context of a **Grothendieck topology**: a category with finite products and a suitable notion of covers (we shall not give the axioms here).

The motivation was to define a cohomology theory that would do for varieties V over fields of prime characteristic what topological cohomology does for varieties over $\mathbb{C}$. The problem is that the Zariski topology is not fine enough (for example, all irreducible curves over a field k are homeomorphic in the Zariski topology) and there is no obvious analogue of the complex topology.

The solution involved the **étale topology**, which is not a topology at all in the classical sense, but is a Grothendieck topology. The analogue of an 'open set' is an étale map to V (see an algebraic geometry text such as Hartshorne [5] for a precise definition, but roughly this is the algebraic analogue of a finite topological covering). The inclusion of a Zariski open set is étale, so the étale topology 'refines' the usual Zariski topology: it is an amazing fact that it is fine enough to allow one to do 'topology' for varieties over a field k of prime characteristic.

We shall not go into any detail here, but for technical reasons étale cohomology (i.e., the cohomology of sheaves in the étale topology) works best for sheaves of torsion groups. For many applications, however, one wants cohomology that takes values in the category of vector spaces over $\mathbb{C}$, or at least over some field of characteristic zero. This can be achieved as follows. Fix a prime number ℓ, and for each natural number n, consider the étale cohomology $H^*(V, \mathbb{Z}/\ell^n\mathbb{Z})$ of V with coefficients in the constant (torsion) sheaf $\mathbb{Z}/\ell^n\mathbb{Z}$. The natural homomorphisms

$$\mathbb{Z}/\ell^{n+1}\mathbb{Z} \longrightarrow \mathbb{Z}/\ell^n\mathbb{Z}$$

induce maps

$$H^*(V, \mathbb{Z}/\ell^{n+1}\mathbb{Z}) \longrightarrow H^*(V, \mathbb{Z}/\ell^n\mathbb{Z}),$$

so we can define $H^*(V, \mathbb{Z}_\ell)$ to be the inverse limit. This is a graded module for the ℓ-adic integers $\mathbb{Z}_\ell$, and so finally we can define the **ℓ-adic cohomology**

$$H^*(V, \mathbb{Q}_\ell) = H^*(V, \mathbb{Z}_\ell) \otimes_{\mathbb{Z}_\ell} \mathbb{Q}_\ell,$$

which is a graded vector space over the field $\mathbb{Q}_\ell$ of ℓ-adic rationals.

For a variety V over a field of characteristic $p \neq \ell$, there are comparison theorems that imply that ℓ-adic cohomology shares many of the familiar properties of the ordinary cohomology of complex varieties.

Since the fundamental work of Deligne and Lusztig constructing representations of finite reductive groups over $\mathbb{Q}_\ell$ using ℓ-adic cohomology, it has become a familiar tool in representation theory. Summaries (designed for representation theorists) of its properties can now be found in several books, such as Carter's [2].

5 Sheaf-theoretic intersection cohomology

We shall now link the sheaf theory of the previous section to intersection homology. Mainly for simplicity, we shall now assume that X is a complex variety of pure complex dimension d, so that for $n = 2d$ we can regard X as an n-dimensional pseudomanifold. Also we shall only use the middle perversity $\overline{m}$: recall that in this case we suppress $\overline{m}$ from the notation, writing $IH_*(X)$ for $IH_*^{\overline{m}}(X)$, and so on.

In fact, although we had not defined a sheaf at that point, we more or less described how piecewise linear Borel-Moore i-chains formed a sheaf. Recall that if $U \subseteq X$ is an open set, then we defined a chain map

$$(C_*^{BM}(X), \partial) \longrightarrow (C_*^{BM}(U), \partial).$$

If we fix i, then

$$U \mapsto C_i^{BM}(U)$$

defines a presheaf C_i^{BM} of complex vector spaces on X, and in fact this is a sheaf.

The sections of this sheaf are of course the Borel-Moore i-chains. A little thought shows that the sections with compact support are just the ordinary i-chains.

Similarly

$$U \mapsto IC_i^{BM}(U)$$

defines a sheaf IC_i^{BM} on X from which the intersection i-chains can be recovered by taking sections with compact support.

The differentials in $(C_*^{BM}(U), \partial)$ and $(IC_*^{BM}(U), \partial)$ induce sheaf homomorphisms

$$C_{i+1}^{BM} \longrightarrow C_i^{BM}$$

and

$$IC_{i+1}^{BM} \longrightarrow IC_i^{BM},$$

giving us chain complexes of sheaves. In fact, it is traditional at this stage to reindex and define $\mathbf{C}_X^{-i} = C_i^{BM}$ and $\mathbf{IC}_X^{-i} = IC_i^{BM}$, so that $\mathbf{C}_X^*$ and $\mathbf{IC}_X^*$ are *cochain* complexes of sheaves on X: i.e., the differential *increases* degree rather than decreasing it. Of course, this is a purely cosmetic change.

We shall consider $\mathbf{C}_X^*$ and $\mathbf{IC}_X^*$ as objects of the bounded derived category $D^b(X) = D^b(\mathrm{Sh}(X))$ of sheaves of complex vector spaces on X. For each i, the sheaves $\mathbf{C}_x^i$ and $\mathbf{IC}_X^*$ are **cohomologically trivial**, meaning that the sheaf cohomology $H^*(X, \mathbf{C}_X^i)$ and $H^*(X, \mathbf{IC}_X^i)$ vanishes in positive degrees: this is a consequence of the fact that they are 'fine' sheaves: see [6, Section 5.2] for more details. Using this cohomological triviality, a standard spectral sequence argument shows that

- $H^{-i}(\mathbf{R}\Gamma(X, \mathbf{C}_X^*)) \cong H_i^{BM}(X)$.
- $H^{-i}(\mathbf{R}\Gamma(X, \mathbf{IC}_X^*)) \cong IH_i^{BM}(X)$.

Thus, for example, $\mathbf{IC}_X^*$, considered up to isomorphism in $D^b(X)$, is enough to recover the Borel-Moore intersection homology of X. This was the key to Goresky and MacPherson's proof that intersection homology is a topological invariant: they exhibited a list of axioms for $\mathbf{IC}_X^*$ that were independent of any choice of stratification and which determined it up to isomorphism in the derived category $D^b(X)$.

Despite this algebraic characterization of $\mathbf{IC}_X^*$, the topological construction was still needed in order to show that there was any object at all that satisfied Goresky and MacPherson's axioms. Then Deligne gave a more algebraic construction that we shall now outline.

First, suppose that

$$S^* = \ldots \longrightarrow S^{p-2} \longrightarrow S^{p-1} \longrightarrow S^p \longrightarrow S^{p+1} \longrightarrow \ldots$$

is a cochain complex of sheaves on X. We shall define a new complex

$$\tau_{\leq p} S^* = \ldots \longrightarrow S^{p-2} \longrightarrow S^{p-1} \longrightarrow K \longrightarrow 0 \longrightarrow \ldots,$$

where K is the kernel of the differential $S^p \longrightarrow S^{p+1}$. Thus there is a natural cochain map

$$\tau_{\leq p} S^* \longrightarrow S^*$$

that induces isomorphisms in cohomology in degrees $i \leq p$, and the cohomology of $\tau_{\leq p} S^*$ is zero in degrees $i > p$. This describes a 'truncation' functor $\tau_{\leq p}$ from the derived category $D^b(X)$ to itself.

Recall that we can define the intersection homology of a complex variety X using a stratification

$$X_0 \subseteq X_2 \subseteq \ldots \subseteq X_{n-2} \subseteq X_n = X$$

that consists of closed subvarieties, so that all the strata have even dimension as pseudomanifolds.

Let

$$i_k : (X - X_{n-2k}) \longrightarrow (X - X_{n-2k-2})$$

be the inclusion map for $1 \leq k < d$, and let i_d be the inclusion map from $(X - X_0)$ to X.

Deligne's construction uses truncation functors and derived functors of direct image functors. We start with the constant sheaf $\mathbb{C}$ on the non-singular part $(X - X_{n-2})$ of X, and then shift it in degree to create the object $\mathbb{C}[n]$ of $D^b(X - X_{n-2})$. Deligne showed that, in $D^b(X)$,

$$\mathbf{IC}_X^* \cong \tau_{\leq -d-1} \mathbf{R}(i_d)_* \ldots \tau_{\leq -n+1} \mathbf{R}(i_2)_* \tau_{\leq -n} \mathbf{R}(i_1)_* \mathbb{C}[n].$$

Of course, this construction still needs a stratification of X, but its big advantage is that it does not need a triangulation. This is vital for extending the definition of intersection cohomology to varieties over fields of positive characteristic: here it is possible to define an algebraic version of the stratification, but the triangulation is a bigger problem. Incorporating the ideas of ℓ-adic cohomology as touched upon in the previous section, and using the construction of Deligne just described, Beilinson, Bernstein and Deligne [1] defined **ℓ-adic intersection cohomology** in positive characteristic $p \neq \ell$.

6 Applications in representation theory

In recent years there have been many and varied applications of intersection cohomology to the representation theory of algebraic groups, mostly either due to Lusztig or inspired by his work. A nice survey is given in Lusztig's 1990 ICM talk [7].

Here we shall give a brief taste of just one of these applications.

For a complex variety X, we have produced a complex $\mathbf{IC}_X^*$ of sheaves on X, from which we can recover the intersection homology of X. Rather than producing homology which takes vector spaces as values, we can produce a sequence of sheaves on X by taking the cohomology of the complex $\mathbf{IC}_X^*$. The **i-th local intersection cohomology sheaf** of X is defined to be the degree i cohomology $\mathbf{H}^i(\mathbf{IC}_X^*)$ of $\mathbf{IC}_X^*$.

Now let G be a semisimple algebraic group over $\mathbb{C}$, let B be a Borel subgroup, and let W be the Weyl group. Then G/B is a projective complex variety, and the Bruhat decomposition expresses G/B as the disjoint union

$$G/B = \bigcup_{w \in W} BwB/B$$

of Bruhat cells. For $w \in W$, let $X_w = BwB/B$ be the Bruhat cell corresponding to w, and let $\overline{X}_w$ be its closure (called a Schubert cell). Then $\overline{X}_w$ is a closed subvariety of G/B which is usually singular, and which is a union of Bruhat cells.

Suppose that for some other element $y \in W$ of the Weyl group, $X_y \subseteq \overline{X}_w$. Then Kazhdan and Lusztig proved that the corresponding Kazhdan-Lusztig polynomial is

$$P_{y,w}(q) = \sum_i a_i q^i,$$

where the coefficient a_i is the dimension of the stalk of the local intersection cohomology sheaf $\mathbf{H}^{2i-n}(\mathbf{IC}_{\overline{X}_w}^*)$ at a point of X_y.

Note that the fact that the coefficients are dimensions of vector spaces implies that they are non-negative integers. This fact is not clear from the original combinatorial definition of the Kazhdan-Lusztig polynomials. See Donkin's lectures for more details on the importance of these polynomials.

References

[1] A. Beilinson, J. Bernstein, P. Deligne, *Faisceaux pervers, Analyse et topologie sur les espaces singuliers (I)*, Astérisque **100** (1982)

[2] R. W. Carter, *Finite groups of Lie type: Conjugacy classes and irreducible characters*, Wiley, New York (1985).

[3] M. Goresky, R. MacPherson, *Intersection homology theory*, Topology **19** (1980), 135–162.

[4] M. Goresky, R. MacPherson, *Intersection homology II*, Invent. Math. **71** (1983), 77–129.

[5] R. Hartshorne, *Algebraic Geometry*, Graduate Texts in Mathematics 52, Springer-Verlag, New York (1977).

[6] F. Kirwan, *An introduction to intersection homology theory*, Pitman Research Notes in Mathematics 187, Longman (1988).

[7] G. Lusztig, *Intersection cohomology methods in representation theory*, in 'Proceedings of the International Congress of Mathematicians, Kyoto 1990', Springer Verlag (1991) 155–174.

An Introduction to the Lusztig Conjecture

Stephen Donkin

School of Mathematical Sciences, Queen Mary and Westfield College, Mile End Rd., London E1 4NS, England

In these lectures we discuss the main and central problem of the representation theory of reductive algebraic groups in characteristic $p > 0$: the problem of determining the formal characters of the irreducible modules. In particular, we discuss the conjecture of Lusztig, [**11**], which predicts the characters of certain key modules from which the character of an arbitrary irreducible module may be determined, provided that p is greater than or equal to $2h - 3$, where h is the Coxeter number of the reductive group. Thanks to work of Kashiwara–Tanisaki, [**4**], Kazhdan–Lusztig, [**6**]-[**10**], Lusztig, [**12**], and Andersen–Jantzen–Soergel, [**1**], the conjecture is now known to hold for $p \gg 0$, in the sense that there is an integer $n(\Phi)$ for each root system Φ such that the conjecture holds for all semisimple, simply connected groups with root system Φ defined over an algebraically closed field of characteristic $p > n(\Phi)$. However, no explicit bound for the integer $n(\Phi)$ is known at the present time (except in a few cases when the rank of Φ is very small). In Lusztig's conjecture, the characters are given as $\mathbb{Z}$-linear combinations of Weyl characters, with coefficients described in terms of the Kazhdan–Lusztig polynomials.

We start in the first section with the general framework. In the second section we go through the example of SL_2. Many of the features of this example are present also in the general set-up and to see this it is convenient to use the Chevalley construction of a semisimple group G over an algebraically closed field K of characteristic $p > 0$, via an admissible $\mathbb{Z}$-form of a finite dimensional module for a complex semisimple Lie algebra. We recall this construction in section 3. At the same time we set out the framework of the representation theory of complex semisimple Lie algebras, for future reference. In section 4 we sketch the proof, by Bernstein, Gel'fand and Gel'fand, of Weyl's character formula. This serves as a model and provides the framework and a source of ideas for assaults on the main problem, as described in section 5. In section 6 we define the Kazhdan–Lusztig polynomials and describe the conjecture of Lusztig. Section 7 is devoted to further aspects of the representation theory of reductive groups, in particular the theory of $(\mathfrak{g}_K, T)$-modules due to Jantzen (where $\mathfrak{g}_K$ is the Lie algebra and T is a maximal torus of the group G over the field K) and contains a very brief description of some aspects of the solution of Lusztig's conjecture for $p \gg 0$. The contribution of Kashiwara–Tanisaki is concerned with the representation theory of affine Kač–Moody Lie algebras. The contributions of Kazhdan–Lusztig and Lusztig are concerned with the connection between the representation theory of affine Kač–Moody Lie algebras and representations of quantum groups over $\mathbb{C}$ at roots of unity. We shall say no more about these aspects but instead make some remarks about the work of Andersen–Jantzen–Soergel, [**1**]. The point here is to show that the representation theory of a semisimple group with given root system "stabilizes" when p is sufficiently large and that this stable value is the "same" as the representation theory of the associated quantum group over $\mathbb{C}$ at a pth root of unity. We concentrate on the algebraic group aspect of their work. Even so, section 7 covers only part of the introduction to [**1**]. The emphasis in presentation on the algebraic group aspects of the overall picture is of course dictated by the author's own narrow area of competence (but nevertheless seems appropriate within this conference programme).

The background material on the structure and representation theory of complex semisimple Lie algebras is taken from Humphreys, [**2**]. Most of the background material on the structure and representation theory of reductive algebraic groups is taken from Jantzen, [**3**]. For further details of the construction and properties of Chevalley groups see Steinberg, [**13**].

1 The general framework

For the most part we shall adopt the notation of Roger Carter's lectures.

Let p be a prime number and let K be the algebraic closure of the field of order p. By an algebraic group over K we mean a linear algebraic group over K. Let G be an algebraic group over K. A finite dimensional (left) KG-module V is called *rational* if for some (and hence every) choice of basis $v_1, \cdots, v_n$, the corresponding matrix representation $\pi : G \to \mathrm{GL}_n(K)$ is a homomorphism of algebraic groups. By a G-module, we mean a finite dimensional rational KG-module.

We fix a maximal torus T of G. We denote by $X(T)$ the character group (or weight lattice) of T and regard $X(T)$ as an additive abelian group. Thus $X(T)$ is isomorphic to $\mathbb{Z}^l$, where l is the rank of G, i.e. the dimension of T. For a G-module V and $\lambda \in X(T)$ we put

$$V^\lambda = \{v \in V \mid tv = \lambda(t)v \text{ for all } t \in T\}$$

and call V^λ the λ-*weight space* of V. We say that $\lambda \in X(T)$ is a weight of V if $V^\lambda \neq 0$. We have $V = \bigoplus_{\lambda \in X(T)} V^\lambda$. We write $\mathbb{Z}X(T)$ for the integral group ring of $X(T)$; this has $\mathbb{Z}$-basis $e(\lambda)$, $\lambda \in X(T)$, and multiplication given by $e(\lambda)e(\mu) = e(\lambda + \mu)$. The formal character $\operatorname{ch} V$ of V is defined to be the element $\sum_{\lambda \in X(T)} \dim V^\lambda \, e(\lambda)$ of $\mathbb{Z}X(T)$.

We now suppose that G is connected and reductive. We let Φ be the set of roots of G with respect to T, i.e. the set of non-zero weights of the Lie algebra $\mathrm{Lie}(G)$ of G (regarded as a G-module via the adjoint action). Let B be a Borel subgroup of G containing T. We write Φ^- for the set of non-zero weights of $\mathrm{Lie}(B)$ and put $\Phi^+ = \{-\alpha \mid \alpha \in \Phi^-\}$. Then Φ is the disjoint union of Φ^+ and Φ^-. We call the elements of Φ^+ (resp. Φ^-) the *positive roots* (resp. *negative roots*) and will often write $\alpha > 0$ (resp. $\alpha < 0$) instead of $\alpha \in \Phi^+$ (resp. $\alpha \in \Phi^-$). For the choice Φ^+ of positive roots, B is the negative Borel subgroup.

Let $W = N(T)/T$ be the Weyl group (where $N(T)$ is the normalizer of T). Then $N(T)$ acts on $X(T)$ by $(n.\lambda)(t) = \lambda(n^{-1}tn)$, for $n \in N(T), \lambda \in X(T), t \in T$. Moreover T is in the kernel of this action and so we have an action of W on $X(T)$. We choose a positive definite, W-invariant, symmetric bilinear form $(\ ,\ )$ on $E = \mathbb{R} \otimes_{\mathbb{Z}} X(T)$ and identify X with a subset of E in the natural manner. For $\alpha \in \Phi$ we put $\check{\alpha} = 2\alpha/(\alpha, \alpha)$. An element $\lambda \in X(T)$ is a *dominant weight* if $(\lambda, \check{\alpha}) \geq 0$ for all $\alpha \in \Phi^+$. We write $X^+(T)$ for the set of dominant weights.

We introduce a partial order $\leq$ on $X(T)$ by decreeing that $\lambda \leq \mu$ if $\mu - \lambda$ is a sum of positive roots. For each dominant weight there is an irreducible G-module $L(\lambda)$ such that the λ weight space has dimension 1 and such that all weights of $L(\lambda)$ are less than or equal to λ. Moreover $\{L(\lambda) \mid \lambda \in X^+(T)\}$ is a complete set of pairwise non-isomorphic simple G-modules. It follows that G-modules V, V' have the same composition factors (counting multiplicities) if and only if $\operatorname{ch} V = \operatorname{ch} V'$. Thus the formal character plays the same role in the representation theory of connected reductive groups that the Brauer character plays in the modular representation theory of finite groups. The character of an irreducible module for a connected reductive group over an algebraically closed field of characteristic zero is given by the famous character formula of H. Weyl. The main problem in the representation theory of connected reductive groups is, in effect, to find a characteristic p version of Weyl's formula.

Problem Find an explicit formula for $\operatorname{ch} L(\lambda)$, $\lambda \in X^+(T)$.

2 An Example

We take $G = \mathrm{SL}_2(K)$. We take T to be the group of diagonal matrices in G and take B to be the group of lower triangular matrices in G. Then $X(T) = \mathbb{Z}\rho$, where $\rho\begin{pmatrix} t & 0 \\ 0 & t^{-1} \end{pmatrix} = t$, for $0 \neq t \in K$. Moreover, we have $\Phi^+ = \{\alpha\}$, where $\alpha = 2\rho$. Let V be the natural module, i.e. $V = \{\begin{pmatrix} a \\ b \end{pmatrix} \mid a, b \in K\}$ on which G acts (on the left) by matrix multiplication. Thus V has basis $v_1 = \begin{pmatrix} 1 \\ 0 \end{pmatrix}$, $v_2 = \begin{pmatrix} 0 \\ 1 \end{pmatrix}$. We form the symmetric algebra $S(V) = K[v_1, v_2]$, on which G acts by algebra automorphisms. Note that the usual grading $S(V) = \bigoplus_{r=0}^{\infty} S^r V$, is a G-module decomposition. Here $S^r V$ is the K-span of $v_1^r, v_1^{r-1}v_2, \ldots, v_2^r$, and has dimension $r+1$. Now T acts on $v_1^i v_2^j$ with weight $(i-j)\rho$. Thus $S^r V$ has highest weight $r\rho$ and it is easy to check that $S^r V$ is irreducible for $0 \le r < p$. Hence we have $L(r\rho) = S^r V$, for $0 \le r < p$.

Note however that $S^p V$ is not irreducible. For $g = \begin{pmatrix} a & b \\ c & d \end{pmatrix}$, we have $gv_1^p = (gv_1)^p = (av_1 + bv_2)^p = a^p v_1^p + b^p v_2^p \in Kv_1^p + Kv_2^p$. We similarly get $gv_2^p \in Kv_1^p + Kv_2^p$, so that $Kv_1^p + Kv_2^p$ is a proper submodule of $S^p V$.

In order to construct all the irreducible G-modules, from our limited supply of irreducible symmetric powers, we need the Frobenius morphism. We define the Frobenius morphism $F : G \to G$ by $F\begin{pmatrix} a & b \\ c & d \end{pmatrix} = \begin{pmatrix} a^p & b^p \\ c^p & d^p \end{pmatrix}$. This is a morphism of algebraic groups. Given a G-module M on which G acts via the representation $\pi : G \to \mathrm{GL}(M)$, we form a new G-module denoted M^F (and called the Frobenius twist of M) which has the same underlying space M but on which G acts via the representation $\pi \circ F : G \to \mathrm{GL}(M)$. Since F is surjective we get that M^F is irreducible whenever M is. But in fact a much stronger statement is true. Let r be a non-negative integer, which we write out in base p as $r = r_0 + r_1 p + \cdots + r_m p^m$ (where $0 \le r_0, \ldots, r_m < p$). Then we have $L(r\rho) = S^{r_0} V \otimes (S^{r_1} V)^F \otimes \cdots \otimes (S^{r_m} V)^{F^m}$.

The above construction is due to Brauer, who also showed that the modules $L(r\rho)|_{\mathrm{SL}_2(q)}$, $0 \le r < q$, form a complete set of pairwise non-isomorphic irreducible $K\mathrm{SL}_2(q)$-modules, where q is a power of p.

3 The Chevalley Construction

The general connected reductive group is the product of a central torus and a semisimple group. Using this structure it is easy to make a reduction, in the main problem, to the case in which G is semisimple. Furthermore, if G is a semisimple group (over K) then there a semisimple, simply connected group $\tilde{G}$ and a surjection of algebraic groups $\tilde{G} \to G$, with finite central kernel. Every module for G is thus naturally a module for $\tilde{G}$ and therefore we may as well take G to be semisimple and simply connected. We shall use Chevalley's construction of such groups. This construction has many excellent features. It gives a very concrete description of G and the structure of the group (maximal torus, Borel subgroup, Weyl group etc.) which plays a prominent role in the main problem. There is a Frobenius morphism, seen in section 2 for $\mathrm{SL}_2(K)$, available in the general context and this is clearly visible via the Chevalley construction. Moreover, many base change arguments in representation theory (from characteristic 0 to characteristic p) are facilitated by this construction.

Let $\mathfrak{g}$ be a semisimple finite dimensional complex Lie algebra. Let $\mathfrak{h}$ be a Cartan subalgebra of $\mathfrak{g}$. For an $\mathfrak{h}$-module V and $\lambda \in \mathfrak{h}^* = \mathrm{Hom}_{\mathbb{C}}(\mathfrak{h}, \mathbb{C})$ we write V^λ for the λ-weight space, i.e.

$$V^\lambda = \{v \in V \mid Hv = \lambda(H)v \text{ for all } H \in \mathfrak{h}\}.$$

Then $\mathfrak{g}$ decomposes, under the adjoint action of $\mathfrak{h}$, as a direct sum of weight spaces. We have $\mathfrak{g}^0 = \mathfrak{h}$ and $\mathfrak{g} = \mathfrak{h} \bigoplus (\bigoplus_{\alpha \in \Phi} \mathfrak{g}^\alpha)$, with $\mathfrak{g}^\alpha$ one dimensional for $\alpha \in \Phi$. Let $\mathfrak{b}^+$ be a Borel subalgebra of $\mathfrak{g}$ containing $\mathfrak{h}$. Then we have $\mathfrak{b}^+ = \mathfrak{h} \bigoplus (\bigoplus_{\alpha \in \Phi^+} \mathfrak{g}^\alpha)$ for a subset Φ^+ (the set of positive roots) of Φ. Moreover we have the triangular decomposition $\mathfrak{g} = \mathfrak{n}^+ \bigoplus \mathfrak{h} \bigoplus \mathfrak{n}^-$, where $\mathfrak{n}^+ = \bigoplus_{\alpha \in \Phi^+} \mathfrak{g}^\alpha$ and $\mathfrak{n}^- = \bigoplus_{\alpha \in \Phi^-} \mathfrak{g}^\alpha$ are nilpotent subalgebras of $\mathfrak{g}$ (and where $\Phi^- = \{-\alpha \mid \alpha \in \Phi^+\}$).

We denote the enveloping algebra of a complex Lie algebra $\mathfrak{f}$ by $U(\mathfrak{f})$. The category of (left) Lie modules for $\mathfrak{f}$ is naturally equivalent to the category of (left) modules for the associative algebra $U(\mathfrak{f})$, and we shall pass freely between these structures. Let $\lambda \in \mathfrak{h}^*$. We write $\mathbb{C}_\lambda$ for $\mathbb{C}$ viewed as a one dimensional $\mathfrak{b}^+$-module via the action $X_\alpha v = 0$, $Hv = \lambda(H)v$, for $\alpha > 0$, $H \in \mathfrak{h}$, $v \in \mathbb{C}_\lambda$. Multiplication $U(\mathfrak{n}^-) \otimes_{\mathbb{C}} U(\mathfrak{b}^+) \to U(\mathfrak{g})$ is a $\mathbb{C}$-space isomorphism so that the induced module $Z(\lambda) = U(\mathfrak{g}) \otimes_{U(\mathfrak{b}^+)} \mathbb{C}_\lambda$, called the *Verma module* labelled by λ, is freely generated over $U(\mathfrak{n}^-)$ by $v^+ = 1 \otimes 1$. Hence $Z(\lambda)$ has $\mathbb{C}$-basis

$$X_{-\alpha_1}^{m_1} \ldots X_{-\alpha_N}^{m_N} v^+, \qquad m_1, \ldots, m_N \geq 0,$$

where $\Phi^+ = \{\alpha_1, \ldots, \alpha_N\}$ (and N is the number of positive roots). Now $X_{-\alpha_1}^{m_1} \ldots X_{-\alpha_N}^{m_N} v^+$ is a weight vector of weight $\lambda - \sum_i m_i \alpha_i$, and it follows that the formal character of $Z(\lambda)$ is

$$e(\lambda) / \prod_{\alpha > 0} (1 - e(-\alpha)) = e(\lambda + \rho) / \prod_{\alpha > 0} (e(\alpha/2) - e(-\alpha/2))$$

where ρ is half the sum of the positive roots. (We refer the reader to Humphreys, [**2**; section 24], for the justification of formal manipulations of characters of infinite dimensional $\mathfrak{g}$-modules.)

Every proper submodule of $Z(\lambda)$ lies in $\bigoplus_{\mu < \lambda} Z(\lambda)^\mu$, so the $Z(\lambda)$ has a unique maximal submodule $M(\lambda)$, say, and the quotient $V(\lambda) = Z(\lambda)/M(\lambda)$ is a simple module of highest weight λ.

The Killing form $(\,,\,)$ on $\mathfrak{h}$ is non-singular so we have, for each $\lambda \in \mathfrak{h}^*$, a unique element $H'_\lambda \in \mathfrak{h}$ such that $(H'_\lambda, H) = \lambda(H)$, for all $H \in \mathfrak{h}$. We transfer the Killing form from $\mathfrak{h}$ to $\mathfrak{h}^*$ by the formula $(\lambda, \mu) = (H'_\lambda, H'_\mu)$, for $\lambda, \mu \in \mathfrak{h}^*$.

For $\alpha \in \Phi$ we put $H_\alpha = 2H'_\alpha/(\alpha,\alpha)$. An element $\lambda \in \mathfrak{h}^*$ is *integral* (and λ is called an integral weight) if $\lambda(H_\alpha) \in \mathbb{Z}$ for all $\alpha \in \Phi$ and λ is called *dominant* if it is integral and we have $\lambda(H_\alpha) \geq 0$ for all $\alpha \in \Phi^+$. For $\alpha \in \Phi$ we put $\check{\alpha} = 2\alpha/(\alpha,\alpha)$. Thus $\lambda \in \mathfrak{h}^*$ is integral if $(\lambda,\check{\alpha}) \in \mathbb{Z}$ for all $\alpha \in \Phi$ and is dominant if it is integral and $(\lambda,\check{\alpha}) \geq 0$, for all $\alpha \in \Phi^+$. We write X for the lattice of integral weights and write X^+ for the set of dominant weights. Let $\Delta = \{\alpha_1,\dots,\alpha_l\}$ be the set of simple roots and let $\omega_1,\dots,\omega_l$ be the fundamental dominant weights, defined by the formulas $(\omega_i,\check{\alpha}_j) = \delta_{ij}$, for $1 \leq i,j \leq l$. Then X is the group of $\mathbb{Z}$-linear combinations of $\omega_1,\dots,\omega_l$ and X^+ is the set of elements of the form $a_1\omega_1 + \cdots + a_l\omega_l$, with $a_1,\dots,a_l$ non-negative integers. For $\lambda \in \mathfrak{h}^*$, the module $V(\lambda) = Z(\lambda)/M(\lambda)$ is finite dimensional if and only if λ is dominant. Every finite dimensional $\mathfrak{g}$-module is completely reducible and $\{V(\lambda) \mid \lambda \in X^+\}$ is a complete set of pairwise non-isomorphic irreducible modules. For a finite dimensional $\mathfrak{g}$-module V we have $V = \bigoplus_{\lambda\in X} V^\lambda$. We write $\mathbb{Z}X$ for the integral group ring of X, with canonical basis $\{e(\lambda) \mid \lambda \in X\}$ and define the character $\operatorname{ch} V$, of a finite dimensional $\mathfrak{g}$-module V by $\operatorname{ch} V = \sum_{\lambda\in X} \dim V^\lambda\, e(\lambda)$. Finite dimensional $\mathfrak{g}$-modules V, V' are isomorphic if and only if $\operatorname{ch} V = \operatorname{ch} V'$.

Let E be the real vector space spanned by X. For $\alpha \in \Phi$ we have the reflection $s_\alpha \in \mathrm{GL}(E)$ defined by $s_\alpha(x) = x - (x,\check{\alpha})x$, $x \in E$. The Weyl group W is the (finite) subgroup of $\mathrm{GL}(E)$ generated by the reflections $\{s_\alpha \mid \alpha \in \Phi\}$. Note that W preserves X and indeed we usually regard W as a group of automorphisms of X. The Weyl group acts on X hence on $\mathbb{Z}X$ and we have $\operatorname{ch} V \in (\mathbb{Z}X)^W$, for every finite dimensional $\mathfrak{g}$-module V. The character of $V(\lambda)$ is given by the famous formula of H. Weyl:

$$\operatorname{ch} V(\lambda) = \sum_{w\in W} \varepsilon(w)e(w\cdot\lambda)/\prod_{\alpha>0}(1-e(-\alpha))$$

where we are using the "dot" action $w\cdot\lambda = w(\lambda+\rho)-\rho$, and $\varepsilon(w)$ is the sign of w, i.e. the determinant of $w \in W$ regarded as a linear transformation of $\mathfrak{h}^*$, and where ρ is half the sum of the positive roots.

Chevalley has shown that it is possible to choose elements $0 \neq X_\alpha \in \mathfrak{g}^\alpha$, for $\alpha \in \Phi$, such that :

(1) $[X_\alpha, X_{-\alpha}] = H_\alpha$, for $\alpha \in \Phi$;

(2) $[X_\alpha, X_\beta] = \pm(r(\alpha,\beta)+1)X_{\alpha+\beta}$, for $\alpha,\beta \in \Phi$ with $\alpha+\beta \in \Phi$, where $r(\alpha,\beta)$ is the least integer such that $\alpha - r(\alpha,\beta)\beta \in \Phi$.

The basis $\{H_\alpha\}_{\alpha\in\Delta} \cup \{X_\beta\}_{\beta\in\Phi}$ is called a *Chevalley basis* of $\mathfrak{g}$. We fix a Chevalley basis. Let V be a finite dimensional $\mathfrak{g}$-module. A Chevalley $\mathbb{Z}$-form is a full sublattice $V_\mathbb{Z}$ (i.e. $V_\mathbb{Z}$ is the $\mathbb{Z}$-span of some $\mathbb{C}$-basis of V) such that $(X_\alpha^a/a!)V_\mathbb{Z} \subseteq V_\mathbb{Z}$, for all $\alpha \in \Phi$ and $a > 0$. We set $V_K = K \otimes_\mathbb{Z} V_\mathbb{Z}$. Note that, for $\alpha \in \Phi$ and $\lambda \in X$, we have $X_\alpha V^\lambda \leq V^{\lambda+\alpha}$ and, since V has only finitely many weights, we get that $X_\alpha^j V = 0$, for $j \gg 0$. Thus, for $t \in K$, we get a K-linear endomorphism $x_\alpha(t) = x_{\alpha,V_\mathbb{Z},K}(t)$ of V_K, such that

$$x_\alpha(t)(1 \otimes v) = \sum_{j \geq 0} t^j \otimes \frac{X_\alpha^j}{j!} v \quad \text{for } v \in V_\mathbb{Z}.$$

Note that $x_\alpha(0)$ is the identity map and that $x_\alpha(s)x_\alpha(t) = x_\alpha(s+t)$, for $s, t \in K$, so that $x_\alpha(t) \in \mathrm{GL}(V_K)$. The Chevalley group $G = G(V_\mathbb{Z}, K)$, is the subgroup of $\mathrm{GL}(V_K)$ generated by $x_\alpha(t)$, as α varies over all roots and t varies over K. We choose a finite dimensional $\mathfrak{g}$-module V whose weights span the weight lattice (as a $\mathbb{Z}$-module) and let $V_\mathbb{Z}$ be a Chevalley $\mathbb{Z}$-form of V. The corresponding Chevalley group $G(V_\mathbb{Z}, K)$ is called *universal.* The justification for this term is the following. Suppose that $V'_\mathbb{Z}$ is any Chevalley $\mathbb{Z}$-form in any finite dimensional $\mathfrak{g}$-module V. Then there is a (unique) group homomorphism $\phi : G = G(V_\mathbb{Z}, K) \to G(V'_\mathbb{Z}, K)$ such that $\phi(x_{\alpha,V_\mathbb{Z},t}) = x_{\alpha,V'_\mathbb{Z},t}$, for all $\alpha \in \Phi$, $t \in K$.

Now V'_K is a $G(V'_\mathbb{Z}, K)$-module and hence a G-module via the map $\phi : G \to G(V'_\mathbb{Z}, K)$. Moreover G is simply connected, with root system Φ, and every semisimple, simply connected, algebraic group over K arises as a universal Chevalley group.

We take V' to be $V(\lambda)$, $0 \neq v^+ \in V(\lambda)^\lambda$ and take $V'_\mathbb{Z} = V(\lambda)_\mathbb{Z}$, the Chevalley lattice generated by v^+. The character of $V(\lambda)_K$ is given by Weyl's character formula and $V(\lambda)_K$ has unique simple quotient $L(\lambda)$. Thus we have

$$\mathrm{ch}\, V(\lambda)_K = \mathrm{ch}\, L(\lambda) + \text{ (a sum of ch } L(\mu)\text{'s with } \mu < \lambda).$$

It follows that there are unique integers $a_{\lambda\mu}$ such that

$$\mathrm{ch}\, L(\lambda) = \sum_{\mu \in X(T)^+} a_{\mu\lambda} \mathrm{ch}\, V(\mu)_K$$

with $a_{\lambda\lambda} = 1$ and $\lambda \geq \mu$ whenever $a_{\mu\lambda} \neq 0$.

Thus the main problem is to determine the integers $a_{\lambda\mu}$, and these integers are predicted by Lusztig's conjecture.

4 Weyl's Character Formula

A very similar situation to that discussed above occurs in the proof by Bernstein, Gel'fand, Gel'fand, of Weyl's character formula (in the form due to Kostant, from which other forms are easily deduced). We look at this proof now to get some ideas of possible ways of proceeding in the algebraic group context.

Let $a \in \mathrm{Cent}(U(\mathfrak{g}))$ (the centre of $U(\mathfrak{g})$). Let v^+ be a highest weight vector of $Z(\lambda)$. Then we have $av^+ = cv^+$ for some $c \in \mathbb{C}$ (since the λ-weight space of $Z(\lambda)$ is one dimensional) and $\{x \in Z(\lambda) \mid ax = cx\}$ is a

$\mathfrak{g}$-submodule of $Z(\lambda)$ containing v^+, and is hence equal to $Z(\lambda)$. Thus we get an algebra homomorphism $\chi_\lambda : \mathrm{Cent}(U(\mathfrak{g})) \to \mathbb{C}$ defined by $av = \chi_\lambda(a)v$, for all $v \in Z(\lambda)$. Now if $V(\lambda)$ is a composition factor of $Z(\mu)$ then $\mathrm{Cent}(U(\mathfrak{g}))$ acts on $V(\lambda)$ via χ_μ and also via χ_λ, and so we get $\chi_\lambda = \chi_\mu$. Now one obtains, via the Harish-Chandra isomorphism $\mathrm{Cent}(U(\mathfrak{g})) \to U(\mathfrak{h})^W$, that $\chi_\lambda = \chi_\mu$ if and only if $\mu = w \cdot \lambda$, for some $w \in W$, where $w \cdot \lambda = w(\lambda + \rho) - \rho$. Hence we get

$$\mathrm{ch}\, Z(\lambda) = \mathrm{ch}\, V(\lambda) + \text{ (a sum of } \mathrm{ch}\, V(\mu)\text{'s with } \mu < \lambda \text{ and } \mu \in W \cdot \lambda).$$

It follows that there is some formula

$$\mathrm{ch}\, V(\lambda) = \sum_{\mu \in W \cdot \lambda} a_{\mu\lambda} \mathrm{ch}\, Z(\mu)$$

with $a_{\lambda\lambda} = 1$ and $\lambda \geq \mu$ whenever $a_{\mu\lambda} \neq 0$. We now fix $\lambda \in X^+$. We write $a(w)$ for $a_{w \cdot \lambda, \lambda}$ so we have

$$\mathrm{ch}\, V(\lambda) = \sum_{w \in W} a(w) e(w(\lambda + \rho)) / \prod_{\alpha > 0} (e(\alpha/2) - e(-\alpha/2)).$$

Thus we have $\psi.\mathrm{ch}\, V(\lambda) = \sum_w a(w) e(w(\lambda + \rho))$, where

$$\psi = \prod_{\alpha > 0} (e(\alpha/2) - e(-\alpha/2)).$$

But the character of $V(\lambda)$ is invariant under the action of the Weyl group and, for $y \in W$, we have $y\psi = \varepsilon(y)\psi$. Thus we get

$$\varepsilon(y)\psi.\mathrm{ch}\, V(\lambda) = \sum_{w \in W} a(w) e(yw(\lambda + \rho)) = \varepsilon(y) \sum_w a(w) e(w(\lambda + \rho)).$$

However, we have $a(1) = 1$ so equating coefficients of $e(\lambda + \rho)$ gives $a(y) = \varepsilon(y)$. Hence we have

$$\mathrm{ch}\, V(\lambda) = \sum_{w \in W} \varepsilon(w) \mathrm{ch}\, Z(w \cdot \lambda).$$

5 Some Fundamental Results

Recall that we have constructed the semisimple, simply connected algebraic group G as the universal Chevalley group $G(V_{\mathbb{Z}}, K)$. Let $v_1, \ldots, v_n \in V_{\mathbb{Z}}$ be a $\mathbb{Z}$-basis of $\mathfrak{h}$ weight vectors and represent $G \leq \mathrm{GL}(V_K)$ as matrices with respect to the basis $1 \otimes v_1, \ldots, 1 \otimes v_n$ of V_K. Then the group T of elements of G represented by a diagonal matrix is a maximal torus in G. The Frobenius map $F : \mathrm{GL}_n(K) \to \mathrm{GL}_n(K)$ induces the Frobenius morphism (also denoted F) on G and the fixed point subgroup $G^F = G(q)$ is $G \cap \mathrm{GL}_n(q)$. For $m \geq 1$ we put

$$X(T)_m = \{a_1\omega_1 + \cdots + a_r\omega_l \mid 0 \leq a_1, \ldots, a_r < p^m\} \subset X^+(T).$$

By writing out the base p expansion of the coefficients a_i we get the "p-adic" expansion $\lambda = \lambda_0 + p\lambda_1 + \cdots + p^r\lambda_r$, with all $\lambda_i \in X_1$. We have the fundamental result of Steinberg describing an arbitrary irreducible G-module in terms of the set of restricted irreducible modules $\{L(\mu) \mid \mu \in X(T)_1\}$.

Steinberg's tensor product theorem $L(\lambda) \cong L(\lambda_0) \otimes L(\lambda_1)^F \otimes \cdots \otimes L(\lambda_r)^{F^r}$.

We also mention another fundamental result of Steinberg which is of great importance in the representation theory of finite Chevalley groups.

Theorem $\{L(\lambda) \mid \lambda \in X(T)_m\}$ *is a full set of irreducible* $KG(q)$*-modules, where* $q = p^m$.

These results are generalisations of properties encountered already in the representation theory of $\mathrm{SL}_2(K)$.

For $\psi = \sum_\mu a_\mu e(\mu) \in \mathbb{Z}X(T)$, we define $\psi^F = \sum_\mu a_\mu e(p\mu)$. Then we have $\mathrm{ch}\, M^F = (\mathrm{ch}\, M)^F$, for a rational G-module M, and from Steinberg's tensor product theorem we obtain

$$\mathrm{ch}\, L(\lambda) = \mathrm{ch}\, L(\lambda_0).(\mathrm{ch}\, L(\lambda_1))^F \ldots (\mathrm{ch}\, L(\lambda_r))^{F^r}.$$

The main problem is thus reduced to the problem of finding $\mathrm{ch}\, L(\mu)$, for $\mu \in X(T)_1$.

We write $U_\mathbb{Z}$ for the subring of the enveloping algebra $U(\mathfrak{g})$ generated by $\{X_\alpha^a/a! \mid \alpha \in \Phi, a > 0\}$. Then $U_\mathbb{Z}$ is a free $\mathbb{Z}$-module and the natural map $\mathbb{C} \otimes_\mathbb{Z} U_\mathbb{Z} \to U(\mathfrak{g})$ is a $\mathbb{C}$-algebra isomorphism. The ring $U_\mathbb{Z}$ is known as the Kostant $\mathbb{Z}$-form of $U(\mathfrak{g})$. A Chevalley $\mathbb{Z}$-form of a finite dimensional $\mathfrak{g}$-module V is then a full $\mathbb{Z}$-lattice $V_\mathbb{Z}$ in V which is $U_\mathbb{Z}$-stable, i.e. satisfies $U_\mathbb{Z} \cdot V_\mathbb{Z} = V_\mathbb{Z}$. Thus $V_K = K \otimes_\mathbb{Z} U_\mathbb{Z}$ is naturally a module for the *hyperalgebra* $U_K = K \otimes_\mathbb{Z} U_\mathbb{Z}$. We have also seen that V_K is naturally a G_K-module. But actually more is true, namely $\mathrm{mod}(G)$ is equivalent to $\mathrm{mod}(U_K)$, the category of finite dimensional U_K-modules. (This is the "so called" Verma Conjecture, proved by J.B. Sullivan and by Cline, Parshall and Scott.)

There is an obvious parallel between certain representations of $U(\mathfrak{g})$ on the one hand and representations of G on the other. The Verma modules $Z(\lambda)$ for $U(\mathfrak{g})$ correspond to the Weyl modules $V(\lambda)_K$ for G (both are universal high weight modules with known character in terms of which one seeks to understand the characters of the irreducible modules). The irreducible modules $V(\lambda)$ for $U(\mathfrak{g})$ correspond to the irreducible modules $L(\lambda)$ for G. Recall, from the previous section, that one has, for $\lambda \in X^+(T)$, the beautiful formula $\mathrm{ch}\, V(\lambda) = \sum_{w \in W} \varepsilon(w) \mathrm{ch}\, Z(w \cdot \lambda)$, which came from proving that $[Z(\lambda) : L(\mu)] \neq 0$ implies $\mu \in W \cdot \lambda$. Fortunately, one has for the algebraic group an analogous result.

Linkage Principle (Humphreys, Kač–Weisfeiler, Carter–Lusztig, Jantzen, Andersen, Doty) *If* $[V(\lambda) : L(\mu)] \neq 0$ *then* $\mu \in W \cdot \lambda + p\mathbb{Z}\Phi$.

It is useful to think about this result in the context of the action of the affine Weyl group. For $a \in X$ let $\tau_a : X \to X$ be translation by a, i.e. the map defined by $\tau_a(x) = a + x$ for all $x \in X$. We write $\mathrm{Aff}(X)$ for the group of all affine transformations of X, i.e. permutations of X of the form $\sigma\tau_a$, where $\sigma \in \mathrm{GL}(X)$ (the automorphism group of the abelian group X) and $a \in X$. We define $\mathcal{W} \leq \mathrm{Aff}(X)$ to be the subgroup consisting of the elements $w\tau_a$, with $w \in W$ and $a \in p\mathbb{Z}\Phi$. Thus $\mathcal{W}$ is the semidirect product $Q \rtimes W$, where $Q = \{\tau_{p\theta} \mid \theta \in \mathbb{Z}\Phi\}$ is the group of translations by p multiples of elements of the root lattice.

Now suppose that the root system Φ is indecomposable and α_0 is the highest short root. Then $\mathcal{W}$ is a Coxeter group with generating set consisting of the elements s_α, $\alpha \in \Delta$, together with $s_{\alpha_0,p}$, where $s_{\alpha_0,p}(x) = x - ((x, \check{\alpha}_0) - p)\alpha_0$, for $x \in X$. We have the "dot" action of $\mathcal{W}$ on $E = \mathbb{R} \otimes_{\mathbb{Z}} X$, given by $w \cdot x = w(x + \rho) - \rho$, $w \in \mathcal{W}, x \in E$. A fundamental domain for this action is the "bottom alcove" $A = \{x \in X \mid 0 < (x + \rho, \check{\alpha}) < p,$ for all $\alpha \in \Phi\}$. In the case A_2, the plane E is broken up under the dot action as follows.

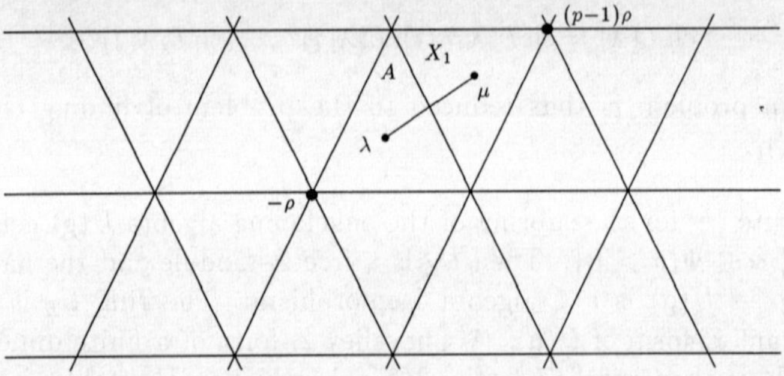

We have $[V(\mu)_K : L(\lambda)] = 1$, $L(\lambda) = \mathrm{ch}\, V(\mu) - \mathrm{ch}\, V(s_{\alpha_0,p} \cdot \mu)$. Here the "dot action" of $\mathcal{W}$ on $\mathbb{R} \otimes_{\mathbb{Z}} X$ is by $w \cdot x = w(x + \rho) - \rho$.

In general we have

$$L((p-1)\rho) = V((p-1)\rho)_K = \mathrm{St}$$

(the Steinberg module), for example by the linkage principle.

Suppose that μ lies inside an alcove. Then we have $\mu = w \cdot \lambda$, for a unique element $\lambda \in A$ and unique element $w \in \mathcal{W}$.

We have

$$\mathrm{ch}\, V(\lambda)_K = \mathrm{ch}\, L(\lambda) + \text{ (a sum of ch } L(y \cdot \lambda)\text{'s with } y \cdot \lambda < w \cdot \lambda)$$

and it follows that there exists an expression

$$\mathrm{ch}\, L(w \cdot \lambda) = \sum_{y \cdot \lambda \leq w \cdot \lambda} a_{yw} \mathrm{ch}\, V(y \cdot \lambda)_K.$$

Actually, by results of Jantzen on translation functors, one can deduce all such from the case of the principal block (containing $\lambda = 0$), provided that this lies in the bottom alcove. So one needs $p > (\rho, \check{\alpha}_0) = h - 1$ (where h is the Coxeter number), i.e. $p \geq h$.

6 Lusztig's Conjecture

Let (W, S) be an arbitrary Coxeter system. We have the Hecke algebra $\tilde{\mathcal{H}}$ over the free polynomial ring $\mathbb{Z}[q]$. Thus $\tilde{\mathcal{H}}$ has $\mathbb{Z}[q]$ basis $\{T_w \mid w \in W\}$ and the multiplication is determined by the formulas

$$T_w T_{w'} = T_{ww'}, \qquad \text{if } l(ww') = l(w) + l(w').$$

$$(T_s + 1)(T_s - q) = 0.$$

Let $A = \mathbb{Z}[q^{1/2}, q^{-1/2}]$ (so here $q^{1/2}$ is an indeterminate and $q = (q^{1/2})^2$) and $\mathcal{H} = A \otimes_{\mathbb{Z}[q]} \tilde{\mathcal{H}}$. We have a ring involution $^-$ on A such that $\overline{q^{1/2}} = q^{-1/2}$. We define an involution $^-$ on $\mathcal{H}$ by

$$\overline{\sum_w a_w T_w} = \sum_w \bar{a}_w T_{w^{-1}}^{-1}.$$

Then we have the following result of Kazhdan–Lusztig.

Theorem *For each $w \in W$ there is a unique element $C_w \in \mathcal{H}$ such that*
(1) $\bar{C}_w = C_w$;
(2) $C_w = \sum_{y \leq w} \varepsilon(y)\varepsilon(w) q^{\frac{1}{2}l(w)-l(y)} \bar{P}_{y,w} T_y$
where $P_{y,w} \in \mathbb{Z}[q]$ has degree at most $(l(w) - l(y) - 1)/2$ and $P_{w,w} = 1$.

Suppose that $-w\rho - \rho$ (i.e. $ww_0 \cdot 0$) is dominant and that $(-w\rho, \check{\alpha}_0) \leq p(p - h + 2)$. We can now state:

Lusztig's Conjecture

$$\operatorname{ch} L(-w\rho - \rho) = \sum_{\substack{y \leq w \\ -y\rho - \rho \text{ dominant}}} (-1)^{l(w)-l(y)} P_{y,w}(1) \operatorname{ch} V(-y\rho - \rho)_K.$$

7 Infinitesimal Theory

We want to get the simple modules by working in a category with enough projective modules and in which the characters of the projective modules determine the characters of the simple modules. Central to the work of Andersen–Jantzen–Soergel is the production of suitable projective modules (for algebraic groups and quantum groups) with the property that a decomposition of these modules looks "the same" for all $p \gg 0$, for algebraic groups, and for quantum groups, with given root system.

We have the triangular decomposition

$$\mathfrak{g}_{\mathbb{Z}} = \mathfrak{n}_{\mathbb{Z}}^+ \oplus \mathfrak{h}_{\mathbb{Z}} \oplus \mathfrak{n}_{\mathbb{Z}}^-$$

of the Chevalley $\mathbb{Z}$-form $\mathfrak{g}_{\mathbb{Z}}$ of $\mathfrak{g}$, where $\mathfrak{n}_{\mathbb{Z}}^+ = \bigoplus_{\alpha>0} \mathbb{Z}X_\alpha$, $\mathfrak{g}_{\mathbb{Z}} = \bigoplus_{i=1}^r \mathbb{Z}H_i$ and $\mathfrak{n}_{\mathbb{Z}}^- = \bigoplus_{\alpha<0} \mathbb{Z}X_\alpha$ (here $H_i = H_{\alpha_i}$, for $1 \le i \le r$). Thus we have K-Lie algebras $\mathfrak{g}_K = K \otimes_{\mathbb{Z}} \mathfrak{g}_{\mathbb{Z}}$, $\mathfrak{n}_K^+ = K \otimes_{\mathbb{Z}} \mathfrak{n}_{\mathbb{Z}}^+$, $\mathfrak{n}_K^- = K \otimes_{\mathbb{Z}} \mathfrak{n}_{\mathbb{Z}}^-$ and a triangular decomposition $\mathfrak{g}_K = \mathfrak{n}_K^+ \bigoplus \mathfrak{h}_K \bigoplus \mathfrak{n}_K^-$. We now write simply X_α for $1 \otimes X_\alpha \in U_K$ and H_i for $1 \otimes H_i \in U_K$.

We write $\underline{u}_1$ for the subalgebra of U_K generated by $\{X_\alpha\}_{\alpha\in\Phi} \cup \{H_i\}_{i=1,\ldots,r}$. Then $\underline{u}_1$ is the restricted enveloping algebra of $\mathfrak{g}_K$ (it has basis

$$X_{\alpha_1}^{a_1} \ldots X_{\alpha_N}^{a_N} H_1^{b_1} \ldots H_r^{b_r} X_{-\alpha_1}^{c_1} \ldots X_{-\alpha_N}^{c_N},$$

where $0 \le a_1, \ldots, a_N, b_1, \ldots, b_r, c_1, \ldots, c_N < p$), and $\dim \underline{u}_1 = p^{\dim \mathfrak{g}}$.

Moreover the relationship between the irreducible G-modules and irreducible $\underline{u}_1$-modules is straightforward.

Theorem (Curtis) $\{L(\lambda)|_{\underline{u}_1} \mid \lambda \in X(T)_1\}$ *is a full set of irreducible* $\underline{u}_1$-*modules.*

But unfortunately, in switching attention from U_K to $\underline{u}_1$, we have lost track of the weight spaces, and hence of the character of a U_K-module. For example, if M is a finite dimensional U_K-module, $\mu \in X$ and $p\mu$ is a weight of M then, for $v \in M^{p\mu}$ we have $H_i v = (p\mu)(H_i)v = 0$. The difficulty is overcome by considering the category of modules for $\underline{u}_1$ and T simultaneously. This is the category of $(\mathfrak{g}_K, T)$-modules introduced by Jantzen. A $(\mathfrak{g}_K, T)$-module is a vector space M with a given action of $\underline{u}_1$ and T such that the action of $\mathfrak{h}_K \le \mathfrak{g}_K$ is the action induced by T, i.e. such that $H_i v = \lambda(H_i)v$ for all $v \in M^\lambda$.

Now we construct induced $(\mathfrak{g}_K, T)$-modules. We put $\underline{u}_1(\mathfrak{b}^+) = \langle X_\alpha, H_i \mid 1 \le i \le r, \alpha > 0\rangle$. Thus $\underline{u}_1(\mathfrak{b}^+)$ is the restricted enveloping algebra of $\mathfrak{b}^+$, in particular we have $\dim \underline{u}_1(\mathfrak{b}^+) = p^{\dim \mathfrak{b}^+}$. Given $\lambda \in X$ we make a one dimensional $(\mathfrak{b}_K^+, T)$-module K_λ such that $X_\alpha v = 0$, $tv = \lambda(t)v$, for $v \in K_\lambda$. Then we have the induced module $\hat{Z}_1(\lambda) = \underline{u}_1(\mathfrak{g}) \otimes_{\underline{u}_1(\mathfrak{b}^+)} K_\lambda$, often affectionately known as a *baby Verma module*. The module $\hat{Z}_1(\lambda)$ has a simple head $\hat{L}_1(\lambda)$ and $\{\hat{L}_1(\lambda) \mid \lambda \in X\}$ is a full set of irreducible $(\mathfrak{g}_K, T)$-modules. For $\lambda \in X$ we have a $(\mathfrak{g}_K, T)$-module, denoted $K_{p\lambda}$, on which T acts with weight $p\lambda$ and on which X_α acts as 0, for α a root and hence $\hat{L}_1(p\lambda) = K_{p\lambda}$. For $\lambda \in X$ we write $\lambda = \lambda_1 + p\mu$, with $\lambda_1 \in X_1$. Then we have $\hat{L}_1(\lambda) \cong \hat{L}_1(\lambda_1) \otimes K_{p\mu}$. Thus the irreducible modules $\hat{L}_1(\lambda)$ with $\lambda \in X_1$ determine all irreducible $(\mathfrak{g}_K, T)$-modules.

Now the category of $(\mathfrak{g}_K, T)$-modules contains enough projective and injective modules. For $\lambda \in X$ we write $\hat{Q}_1(\lambda)$ for the projective cover of $\hat{L}_1(\lambda)$. Then $\{\hat{Q}_1(\lambda) \mid \lambda \in X\}$ is a full set of projective indecomposable $(\mathfrak{g}_K, T)$-modules. Moreover, we have $\hat{Q}_1(\lambda + p\mu) \cong \hat{Q}_1(\lambda) \otimes K_{p\mu}$, for $\lambda.\mu \in X$.

By a $\hat{Z}_1$-filtration of a $(\mathfrak{g}_K, T)$-module Q we mean a filtration $Q = Q_0 \geq Q_1 \geq \cdots \geq Q_{m+1} = 0$ such that, for each $1 \leq i \leq m$, the section Q_i/Q_{i+1} is either 0 or isomorphic to $\hat{Z}_1(\mu_i)$, for some $\mu_i \in X$. Since $\hat{Z}_1(\mu)$ has highest weight μ, the cardinality of $\{i \in [1,m] \mid \mu_i = \mu\}$ is independent of the choice of $\hat{Z}_1$-filtration. We write this number $(Q : \hat{Z}_1(\mu))$.

Theorem *For $\lambda \in X$ the module $\hat{Q}_1(\lambda)$ has a $\hat{Z}_1$-filtration and furthermore, we have $(\hat{Q}_1(\lambda) : \hat{Z}_1(\mu)) = [\hat{Z}_1(\mu) : \hat{L}_1(\lambda)]$, for $\mu \in X$.*

It follows that, given the characters of the $\hat{Q}_1(\lambda)$, one can determine the characters of $\hat{L}_1(\mu)$, for $\mu \in X$. Recall that $L((p-1)\rho) = V((p-1))_K$, and by Weyl's dimension formula we have $\dim V((p-1)\rho)_K = p^N$, where $N = |\Phi^+|$. Thus we have $\hat{L}_1((p-1)\rho) = \hat{Z}_1((p-1)\rho)$ and we get also $\hat{Q}_1((p-1)\rho) = \hat{L}_1((p-1)\rho) = \hat{Z}_1((p-1)\rho)) = L((p-1)\rho)$.

For λ arbitrary, the character of $\hat{Q}_1(\lambda)$ will be very complicated. However, we have the easily describable special case (for $p > h$)

$$\operatorname{ch} \hat{Q}_1(w_0 \cdot 0) = \sum_{w \in W} \operatorname{ch} \hat{Z}_1(w \cdot 0)$$

and this is the starting point of the analysis of Andersen–Jantzen–Soergel.

Now take $\hat{Q}_1(w_0 \cdot 0)$ and apply a sequence of "wall crossings" I to get a new projective module Q_I. Now there exists a finite set $\mathcal{S}$ of such I such that the decomposition of Q_I, for all I in $\mathcal{S}$, as a direct sum of $\hat{Q}_1(w \cdot 0)$'s determine all the $\hat{Q}_1(w \cdot 0)$'s.

Theorem (Andersen–Jantzen–Soergel) *There exists a ring $\mathcal{E}$, which is finitely generated over $\mathbb{Z}$, such that the algebra $K \otimes_{\mathbb{Z}} \mathcal{E}$ is isomorphic to $\operatorname{End}_{(\mathfrak{g}_K,T)}(\bigoplus_{I \in \mathcal{S}} Q_I)$. Moreover, we have a decomposition $\mathcal{E} = \bigoplus_{I,J \in \mathcal{S}} \mathcal{E}_{I,J}$ such that the isomorphism takes $K \otimes_{\mathbb{Z}} \mathcal{E}_{I,J}$ to $\operatorname{Hom}_{(\mathfrak{g}_K,T)}(Q_I, Q_J)$.*

However, on general grounds, there is a sense in which a decomposition of 1 in $K \otimes_{\mathbb{Z}} \mathcal{E}$, as an orthogonal sum of primitive idempotents, stabilizes for $p \gg 0$. This gives that $\operatorname{ch} \hat{Q}_1(\lambda)$ is "independent of p" (for $\lambda = w \cdot 0$, $w \in W$). Moreover, this decomposition is the same as in the quantum case giving that $\operatorname{ch} L(w \cdot 0)$ (for $p \gg 0$) is the same as in the quantum case. Since Lusztig's conjecture is known to hold in this case, by work of Kashiwara–Tanisaki, Kazhdan–Lusztig and Lusztig, one is done.

It remains to try to give some idea how one may describe $\operatorname{Hom}(Q_I, Q_J)$ independently of characteristic. In order to do this it is necessary to enlarge the framework of $(\mathfrak{g}_K, T)$-modules somewhat. Let $U(\mathfrak{g}_K)$ be the enveloping algebra of $\mathfrak{g}_K$ and define $\bar{U}(\mathfrak{g}) = U(\mathfrak{g}_K)/\langle X_\alpha^p \mid \alpha \in \Phi \rangle$ then the symmetric algebra $S(\mathfrak{h}_K) = U(\mathfrak{h}_K)$ embeds in $\bar{U}(\mathfrak{g})$. For any $S(\mathfrak{h}_K)$-algebra A we define a category of modules $\mathcal{C}_A$. An object in $\mathcal{C}_A$ is an X-graded A-module $M = \bigoplus_{\lambda \in X} M^\lambda$ with an $\mathfrak{h}_K$-compatibility property (if H_i maps to $t_i \in A$ then we require that $H_i v = t_i v + \lambda(H_i) v$, for $v \in M^\lambda$). Now one needs to make a fresh start and establish analogues for $\mathcal{C}_A$ of the basic properties of the category of $(\mathfrak{g}_K, T)$-modules. We take for A the localization of $S(\mathfrak{h}_K)$ at the ideal generated by $H_1, \ldots, H_r$. We put $A^\emptyset = A[H_\alpha^{-1} \mid \alpha > 0]$ and we put $Q_{A,I}^\emptyset = A^\emptyset \otimes_A Q_{A,I}$, etc. Then each $Q_{A,I}^\emptyset$ is a direct sum of modules $Z_A^\emptyset(w \cdot 0)$ so that $\mathrm{Hom}_{\mathcal{C}_A}(Q_{A,I}^\emptyset, Q_{A,J}^\emptyset)$ is "independent of p". For a positive root β we put $A^\beta = A[H_\alpha \mid \beta \neq \alpha \in \Phi^+]$, for $\beta \in \Phi^+$. We put $Q_{A^\beta,I} = A^\beta \otimes_A Q_{A,I}$. Then we have

$$\mathrm{Hom}_{\mathcal{C}_A}(Q_{A,I}, Q_{A,J}) = \cap_{\beta>0} \mathrm{Hom}_{\mathcal{C}_{A^\beta}}(Q_{A^\beta,I}, Q_{A^\beta,J}).$$

Now each term $\mathrm{Hom}_{\mathcal{C}_{A^\beta}}(Q_{A^\beta,I}, Q_{A^\beta,J})$ is independent of p and with a great deal of work it is shown that the embeddings of $\mathrm{Hom}_{\mathcal{C}_{A^\beta}}(Q_{A^\beta,I}, Q_{A^\beta,J})$ in $\mathrm{Hom}_{\mathcal{C}_A}(Q_{A,I}, Q_{A,J})$ are independent of p and we get the desired conclusion, i.e. that $\mathrm{Hom}_{(\mathfrak{g},T)}(Q_{K,I}, Q_{K,J}) = K \otimes_A \mathrm{Hom}_{\mathcal{C}_A}(Q_{A,I}, Q_{A,J})$.

References

1. H. H. Andersen, J. C. Jantzen and W. Soergel, *Representations of quantum groups at a pth root of unity and of semisimple groups in characteristic p : Independence of p*, Astérisque **220**, Paris 1994 (Soc. Math. de France).

2. J. E. Humphreys, *Lie Algebras and Representation Theory*, Graduate Text in Mathematics **9**, Springer 1972, Berlin/Heidelberg/New York.

3. J. C. Jantzen, *Representations of Algebraic Groups*, Pure and Applied Mathematics, **131**, Academic Press 1987.

4. M. Kashiwara and T. Tanisaki, "Characters of the negative level highest weight modules for affine Lie algebras", *Internat. Math. Res. Notices* **3** (1994), 151–161.

5. D. Kazhdan and G. Lusztig, "Representations of Coxeter groups and Hecke algebras", *Invent. Math.* **53** (1979), 165-184.

6. D. Kazhdan and G. Lusztig, "Affine Lie algebras and quantum groups", *Internat. Res. Notices* **1991**, no. 2, 21-29, in : *Duke Math. J.* **62** (1991).

7. D. Kazhdan and G. Lusztig, "Tensor structures arising from affine Lie algebras I", *J. Amer. Math. Soc.* **6** (1993), 905-947.

8. D. Kazhdan and G. Lusztig, "Tensor structures arising from affine Lie algebras II", *J. Amer. Math. Soc.* **6** (1993), 949-1011.

9. D. Kazhdan and G. Lusztig, "Tensor structures arising from affine Lie algebras III", *J. Amer. Math. Soc.* **7** (1994), 335–381.

10. D. Kazhdan and G. Lusztig, "Tensor structures arising from affine Lie algebras IV", *J. Amer. Math. Soc.* **7** (1994), 383–453.

11. G. Lusztig, "Some problems in the representation theory of finite Chevalley groups," pp.313-317 in B. Cooperstein, G. Mason (eds), *The Santz Cruz Conference on Finite Groups* (1979), Proc. Symp. Pure Math. **37**, Providence R.I. 1980 (Amer. Math. Soc).

12. G. Lusztig, "Monodromic systems on affine flag manifolds, Proc. Roy. Soc. London Ser. A **445** (1994), 231–246. .

13. R. Steinberg, "Lectures on Chevalley Groups," *mimeographed lecture notes*, New Haven : Yale Univ. Math. Dept. 1967.

Index